Plough Music

First published in 2009 by
Liberties Press
Guinness Enterprise Centre | Taylor's Lane | Dublin 8 | Ireland
www.LibertiesPress.com
info@LibertiesPress.com
+353 (1) 415 1286

Trade enquiries to CMD BookSource
55A Spruce Avenue | Stillorgan Industrial Park
Blackrock | County Dublin
Tel: +353 (1) 294 2560 | Fax: +353 (1) 294 2564

Distributed in the United States by
Dufour Editions
Po Box 7 | Chester Springs | Pennsylvania | 19425

and in Australia by
InBooks
3 Narabang Way | Belrose NSW 2085

Copyright © David Medcalf, 2009
The author has asserted his moral rights.
ISBN: 978-1-905483-78-5
A CIP record for this title is available from the British Library.

Cover design by Liberties Press and Ros Murphy
Internal design by Liberties Press
Illustrations by Michael Shannon
Printed by ScandBook

This book is sold subject to the condition that it shall not, by way of trade or otherwise, be lent, resold, hired out or otherwise circulated, without the publisher's prior consent, in any form other than that in which it is published and without a similar condition including this condition being imposed on the subsequent publisher.
No part of this publication may be reproduced or transmitted in any form or by any means, electronic or mechanical, including photocopying, recording or storage in any information or retrieval system, without the prior permission of the publisher in writing.

Plough Music

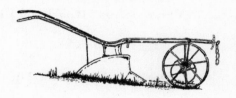

David Medcalf

Contents

	Introduction	9
1	'Quick, Come See!'	11
2	Céide Fields	15
3	When Harry Met Henry	21
4	The Ursus, From Poland with Pride	24
5	Kellystown's Vintage Heroes	27
6	Old Boy Racers	30
7	Plough Music	34
8	Deere John	37
9	The Blessed Trinity	40
10	The Gentleman's Tractor	43
11	Minnows – and Sharks	46
12	The 1994 National Ploughing Championships	52
13	The Psychology of Ploughing	56
14	Making Progress	59
15	Tractor Heaven, USA	63

16	THE CRADLE OF THE WORLD PLOUGHING CHAMPIONSHIPS? ER, WORKINGTON . . .	67
17	THE MAN IN THE MIDDLE	71
18	THE EARLY DAYS OF COMPETITIVE PLOUGHING	75
19	PETER GRAHAM DIED! OH NO HE DIDN'T!	78
20	AN A-TO-Z	82
21	ROMANTIC IRELAND DEAD AND GONE? NOT A BIT OF IT!	86
22	'NO WAY WOULD I GO BACK TO PLOUGHING'	90
23	THE WEATHER MAN	93
24	A MINORITY PURSUIT	97
25	ACRES AND ACRES	100
26	'WE CAN BE HEROES'	104
27	THE ORIGINAL CELTIC FIELDS	108
28	THE DIET OF WORMS	112
29	FULL STEAM BEHIND	116
30	THE FIRST TRACTORS	120
31	DOWN BY THE BRAHMAPUTRA	124
32	THE GEE-GEES	130
33	OFF-ROADERS	134
34	GRUB'S UP!	137
35	NOR PLOW	141
36	A PLANT IN THE WRONG PLACE	145
37	THE WEST AWAKE	149

38	Mud, Mud, Glorious Mud	153
39	They Didn't Lick It Off the Stones	157
40	How to Make a Small Fortune from Farming? Start with a Large One	161
41	The Secrets of the Art	164
42	Emergency Measures	170
43	From Athlone to Palataka	174
44	The Thin Blue Line	178
45	Wide Open Spaces	182
46	Give Up Yer 'Oul Ploughs?	186
	Acknowledgements	191

INTRODUCTION

When Michael Freeman and the folk in Liberties Press came up with the idea of a book on ploughing, they turned to a city boy to do the writing. I hail from Dublin and the house where I live in Enniscorthy does not have so much as a garden, let alone a field of winter wheat out the back.

However, since first arriving to work as a journalist in County Wexford in 1988, I have always been drawn to the farming stories. The lads in the print works at *The People* were hugely amused at my efforts to shear a sheep in the interest of investigative reportage. While they laughed and made 'baa' noises every time I entered the building, I was becoming enamoured with all matters rural.

The intense identification with place, with 'the home place', among country dwellers is enviable in the eyes of someone such as myself who may be vaguely classified as a 'southside' Dub, and an exiled one at that. The awareness of townland and parish lends an intimate drama to the events of country living which is impossible in the whirlpool of urban existence.

As I set about exploring the specific subject of tillage farming, I found plenty of people willing to share their dramas and histories and specialities. There was no one who turned away enquiries – though they came from someone who still has not mastered the etiquette of whether to knock at the front door or the back when calling at farmhouses.

Thanks are hereby expressed to everyone who shared their thoughts and their tea and their cake. Particular gratitude is due to Matt O'Toole, interviewed while in his ninetieth year, who worked as a ploughman behind a brace of horses during his youth. A shrewder, less sentimental observer of the countryside would be hard to find.

Ploughing, tillage and agriculture in general face a future which is worryingly uncertain. There is no guarantee that the system will retain the family farm structure of the past eighty years and more. I am optimistic that the common sense, good humour and spirit of cooperation innate in country dwellers will ensure that something worthwhile emerges from trying times.

DAVID MEDCALF

1

'QUICK, COME SEE!'

> 'Virgil, quick, come see . . .'
> THE BAND

Yer man Bono is only in the ha'penny place. The rock star may have shared platforms with presidents, supped plonk with prime ministers and been slapped on the back by taoisaigh. But when it comes to an artist having constant access to the heart of power, Virgil is hard to beat. And his influence stemmed in no small part from his good standing as an agricultural adviser.

Poet Publius Vergilius Maro was born in the year 70 BC of farming stock in Lombardy. He became an accomplished composer of verse, having gone to Naples to sit as a student at the feet of Siro in the quest to perfect his understanding of the craft.

Like Bono, he appears to have preferred a snappy nickname to the full handle. So it was as Virgil that he went on to become an established member of the entourage that surrounded Octavian — more often referred to as Augustus Caesar, omnipotent head of the Roman Empire.

The versifier from Mantua is best remembered for his epic *The Aeneid*. Written in dactylic hexameter, no less, it is a real page-turner, as it follows the travels and travails of its hero Aeneas. Classical scholars believe that the author spent eleven years writing the work, which won him literary immortality.

The Aeneid remained unfinished when Virgil died in 19 BC while returning to Italy from Greece with Augustus. The poet's will was that *The Aeneid* be destroyed after his death, but this wish was ignored, at the express order of the emperor.

It was a more humdrum accomplishment, however, which originally gave Virgil the leg-up to his position of great influence. The long poem *Georgics* is little read nowadays, two millennia on from when it was standard literary fare. *Georgics* is a farmer's manual, a work of rural inspiration and reference covering everything from the choice of crops to the keeping of beehives.

Granted, Virgil managed to work some politics into the Latin rhyme scheme, along with all the stuff about propagation of trees and choosing the correct quality of seed corn. He had a particular horror of civil war, such as had convulsed his native Po valley, and this is reflected in the poem.

Still, it is possible to read most of *Georgics* as an agricultural handbook, and as such it was widely enjoyed during the poet's lifetime, when it was very well known and had a deep influence. It was as though Seamus Heaney, in the guise of a professor of agricultural science, had become a top government adviser while simultaneously topping the best-seller charts.

Farming and food production (and, indeed, poetry) evidently mattered to the literate classes of the Roman Empire. Virgil was likely to be called upon to give public performances from *Georgics* or to deliver private readings for Augustus and his circle. It is impossible to imagine a latter-day Cabinet being transfixed by the call to the fields delivered by Virgil:

> It's time again to put the bull before the deep-
>   pointed plough to pull his weight
> And have the share glisten, burnished by the
>   broken sod.

Virgil was something of a fair-weather ploughman, suggesting that such work was best undertaken on days when you could happily strip to the waist. Nevertheless, his advice is all well grounded, not alone in his own experience but also in the less high-flown writings of a venerable farmer called Varro.

Varro never achieved the fame of Virgil, just as many a protest singer falls short of the sales and celebrity enjoyed by U2.

> Ploughing hardware improved little from Virgil's time in the first century before Christ up to the sixteenth century, according to Liam Stafford. Around that time, an iron ploughshare was offered as an alternative to the traditional wooden one, though most of the set remained timber. In 1796, Charles Newbold took out a patent on a cast-iron plough in the United States. This was followed in 1837 by John Deere and Leonard Andrus, who turned to steel as their material of choice.

Georgics by Virgil, translated by Peter Fallon, Oxford World Classics 2006

With thanks to Liam Stafford

2

CÉIDE FIELDS

> 'A long, long time ago'
> DON MCLEAN ('AMERICAN PIE')

Hello, my name is David. Lovely place you have here in North Mayo. Really, spectacularly, gob-smackingly lovely.

After travelling the width of Ireland to be with you, I confess feeling just a tad removed from my usual environment. I swear to goodness I haven't clapped eyes on an ear of grain or a hint of a cornfield since the Meath/Westmeath border. That's more than two hundred kilometres of unrelieved grassland countryside. Cattle and sheep only. Divil a bit of wheat or barley.

I honestly do not wish to appear too pass-remarkable. Please don't take offence. Still I must respectfully put it to you that the Emerald Isle lost its green sheen somewhere back there along the road out of Ballina heading for Ballycastle. 'And there's forty shades of brown', as the song might have had it. Now here we are in the Céide Fields above the cliffs near Drumpatrick Head. The word 'wilderness' is being measured at the back of my mind, just to see if it fits.

This is a world of cormorants and heather. This is a world of precious few houses, a world with endless expanses of windswept moor. This is a world away from the fertile fields of the south and east. Yet this is where over thirty thousand people – my humble self included – come every year as tourists of all nationality to explore the birth of Irish agriculture. This remote and ravishingly attractive spot is where it all began, unlikely though this seems. Here are the Céide Fields, complete with their modern, pyramid-shaped interpretive centre set in the side of a marshy hill.

Turn the clock back more than six thousand years ago to find an Ireland which was shrouded in dense forest. The few humans who lived inland on this island were likely of the hunter-gatherer persuasion. There were certainly colonies of fisherfolk too in communities along the coast, including the coast of north Mayo. Then, in later Stone Age times, along came a more sophisticated class of settler. The newcomers hauled down trees to create the space for grazing their own domesticated livestock and, most radically of all, they grew their own grain rich grasses.

It is practically certain that the Irish farmers of around

4000 BC were not confined to North Mayo. The difference is that, elsewhere, traces of their activities have generally been erased beyond recognition through the subsequent scratchings at the soil by succeeding generations of food producers. At the Céide Fields and nearby Belderrig, however, the brown bog advanced to cover their enterprises and preserve them for modern man to scrutinise with wonder and admiration. It was, as archaeologist Séamas Caulfield has pithily observed, a 'slow-motion Pompeii'.

Where the city of Pompeii was smothered under volcanic dust in a catastrophic instant, it has taken five millennia to lay down the five metres of peat under which the most deeply covered fields at Céide lie entombed. The farms of the Stone Age were slowly annexed and the farmers who tended them were more or less forgotten, until a man of genius recognised the signs and began to explore beneath the bog.

It was in 1934 that Belderrig school master Patrick Caulfield noted the pattern of stone walls being uncovered during turf cutting. He wrote to the National Museum of Ireland to apprise them of his observations. The museum authorities in Dublin responded with polite acknowledgement but they were not stimulated into taking any action. It was eventually left to Patrick's son to carry out the necessary excavations and make full sense of the older man's remarkable insights.

Young Séamas had migrated east and followed his father into the teaching profession before transferring into archaelology, in due course taking a position on the staff of University College Dublin. When the college set about exploring a megalithic tomb in his native county of Mayo, he was glad of the excuse to spend time on native turf. While his discoveries there may lack the Hollywood drama of the famous Egyptologists, he experienced his tomb-of-King Tutankhamen equivalent.

It came in the early seventies when Professor Caulfield was supervising the efforts of his neighbour, the late John 'Ban' McCann. John had volunteered to assist in the dig at the Céide Fields. When his spade struck the hard ridges where Neolithic farmers once grew grain, the professor felt the hair rise on the back of his neck with excitement.

'I could hear the spade grating on the stone tops of the ridges. It was one of the great moments of my career.' Here was conclusive evidence of a civilised, peaceable, organised people capable of growing their own corn. Subsequent analysis of pollen samples proved that they cultivated wheat and barley, with oats and rye coming later. Not all of their complex Céide operation has been exposed, and it is unlikely that much more of it will ever be brought out into light. Rather than digging down, the archaeologists have instead used probes pushed into the bog to chart a network of enclosures bounded by boulder walls.

They estimate that this sophisticated cooperative venture spread over four square miles and that a quarter million tonnes of stone were shifted in the process. The principal activity was the rearing of livestock for meat. However, there was scope too for grain in the shelter of those exhaustingly assembled walls. 'Ban' McCann's shovel had hit upon a system of ridges where wheat was cultivated in the same style as potatoes are today.

Eerily, McCann recalled growing oats in ridges as a young man, making him an unwitting inheritor of a tradition that went back thousands of years. The ridges suggest that the grain was hand-reared but the discovery of a stone plough tip at the Céide Fields implies that the inhabitants rose to more intensive arable methods. The wooden frame of the original plough has long since melted into the bog, but the exhibition at the interpretive centre offers a plausible re-creation of what it must have looked like.

The mischievous professor suggests that such activity must have had a strong feminine input. Ancient men liked to boast of their successes on the hunting field. And the size of the stones in the wall suggests that masculine strength was required to lever them into place. However, the gathering of wild grass seed was surely the work of women and they must have led the way in refining the practice by sowing seed close at hand. They were probably doing something similar along the way at Belderrig, though Bronze Age intrusions have muddied the waters there and made it impossible (so far) to put an accurate date on the marks left by ancient ploughs.

Scholars who pick over the details of such things, call the old Neolithic plough an 'ard' – related in its etymology to the word 'earth' – and snootily insist that it disturbed the very top layer of soil only, without slicing a decent sod. The early farmers were lucky if they made it down to a depth of 30 centimetres. The disturbance was so superficial that it was necessary to take a second run at the chosen patch of ground, at right angles to the first pass.

Still, they were mighty men, and mighty women, to carve such a good life for themselves out of the windlashed heights of the Céide Fields.

With thanks to Professor Séamas Caulfield, to Declan Caulfield and to the OPW's staff at Céide Fields

Ancient Farming by Peter J. Reynolds, Shire Publications 1987

3

WHEN HARRY MET HENRY

The children of Ireland should learn about Harry Ferguson at their mother's knee. Industrialist, entrepreneur and motor-racing enthusiast Henry George Ferguson (1884–1960) is a hero of considerably more substance than many people who are much better known.

His greatness as the deviser of the three-point linkage is now acknowledged with a plaque on the wall of a farmhouse in the townland of Growell (pronounced not so much 'growl' and more 'grow well', which has to be appropriate) in County Down, where he was born.

On the other side of the road, across the carriageway from his sturdy grey-walled childhood home, Ferguson devotees have ripped out the ditch to erect a handsome statue of the great man, posed leaning on a farm gate, with spanner in hand.

It's all a load of hokum, of course. Ferguson was not at all a man to waste much time leaning on rustic gates in idle contemplation. In fact, young Harry could scarcely wait to leave Growell and the soggy banks of Loughaghery. He yearned to fly planes or invent fast cars – and indeed he did both. He was, after all, the man behind the Jensen Interceptor, with glamorous shades of James Bond.

Yet his reputation is firmly grounded in the mundane, countryman's ingenuity of the three-point linkage and all that flowed from it in terms of tractor design and manufacture. His genius allowed even a small tractor to make a decent job of ploughing, as the linkage allowed the power of the engine to be channelled into performing the task efficiently and safely.

The notion that Harry Ferguson should be the patron saint of the modern plough appears fanciful when viewed from the roadside shrine to his memory in Growell. His part of County Down is drumlin country. Its small, boggy fields are fit for sheep – and sheep wearing wellingtons, at that. There is probably no land suitable for ploughing within five miles of this rushy, squelchy spot.

Yet it was this restless son of one of Ulster's soggier corners who went from working in his uncle's Belfast bicycle shop to become a significant industrialist by the time he was in his forties. With his innovative designs – groundbreaking in the most literal sense – in his briefcase, he went to the United States to meet the great Henry Ford as an equal.

Two countrymen, they were for all the world like a couple of farmers haggling over the sale of a few bullocks on a fair day. No lawyers were required at Ford's pad in Dearborn, Michagan, when they shook hands in 1938. Yet the stakes were high, and the implications of the deal momentous.

The deal, agreed eyeball to eyeball, spawned millions of tractors on the back of the American's manufacturing muscle and the Ulsterman's ability to focus the power of a tractor's engine where it matters. The partnership lasted only until 1947, when Henry senior died, but by then the two men had delivered machinery which helped to feed two continents in the midst of a cataclysmic war.

When the handshake deal sundered, the American made Cork – at the southern end of Ireland – the principal European bridgehead for his company's farm-machinery operation. Though Ferguson took over $9 million dollars from the split, he lost control of his designs.

Ironically, he found a new partner in Ford's back yard. He allied himself with the Massey Company of Canada, with which his name became forever associated. Nowadays, the world of vintage farm machinery remains polarised between the Ford and Ferguson camps.

Yet even the most fanatic Ford enthusiast has to concede that the man from Growell had no equal when it came to making the most of the output of a small tractor.

4

THE URSUS, FROM POLAND WITH PRIDE

Poles and Irish, compare and contrast: two nations that have done their share of ploughing. Potatoes and Catholicism provide obvious points of common reference. The notable lack of any great traditional cuisine is another parallel. Where one has beetroot soup, the other has pigs' trotters – in either case, scarcely the stuff of gourmet dinners. For both countries, their international repute is based less on food and more on drink.

However, the folk from the banks of the Vistula trump the Irish hands down when it comes to having foreign oppressors to moan about. The uneasy relationship between Ireland and Britain is a mere adolescent affair, laced with more love than is commonly acknowledged, compared with the bloody *ménage à trois* by the Baltic.

In the threesome of Poland, Germany (often in its previous guise of Prussia) and Russia, the first-named has been cast in the role of perpetual doormat beneath the alternating jackboots of its neighbours. No great love is lost there.

The Irish across much of the island have made a great fuss of retaining their own language as a symbol of cultural identity. The reality is that practically everyone is happy to speak the tongue of the colonisers.

In contrast, it is a matter of great pride to practically all Poles that they have clung on to their own language. It remains intact in all its Slavic glory, despite centuries of domination by overweening invaders.

They have every right to be proud of their coal, too, of course. With five hundred universities for a population of thirty-eight million, they may also hold their heads up as an educated people. The fact that there seems to be a college on every street corner goes some way towards explaining how the immigrant on the killing line in the meat factory turns out to be a mathematics graduate, while the builder's labourer is a doctor – either of letters or medicine.

Talk to any of them, and you may be surprised to learn that the Poles are extremely proud of the Ursus; as proud, indeed, as the Irish are of Bailey's or the English of Stevie Gerrard. Ursus. What a ridiculous name for a tractor.

Ursus conjures up thoughts of the Great Bear constellation of stars. While Ursa Major lights up the sky, Ursus Tractor is firmly planted on the ground, scuffing the rich earth which made Poland so popular with Russian tsars and Prussian kings.

Like several other tractor marques — such as Valmet in Finland — the Ursus is an example of swords being turned into ploughshares. Between world wars, the company was more of a military concern than an agri-enterprise, but it rose from the rubble of Warsaw to provide the mechanical workhorses for Polish farming.

Poles are brought up to believe that their Ursus is a triumph of home-grown engineering which bestrides the world from steppe to prairie. Sorry, Krzysztos, but the reality is that the brand has owed much to Zetor and Massey Ferguson technology, while the Ursus's near-monopoly in its own haggard has never extended to the international stage.

With a home market of two million farmers, Ursus continues to produce good, workaday small tractors, reflecting Poland's hard-working small farmers. Nevertheless, the cachet surrounding the marque underlines the fact that every farming nation with its own tractor seems to have a soft spot for tractors.

While the Irish go dewy-eyed over the Ferguson and the Poles wax lyrical about the Ursus, the Belarussians are doubtless besotted with the (you guessed it) Belarus, the Austrians are sentimental about the Steyr, the Italians wax lachrymose about the Landini, and the Hungarians presumably hark back to the heyday of the Hatz.

5

KELLYSTOWN'S VINTAGE HEROES

It has not rained in weeks. Yet somehow, the track is mysteriously ankle-deep in mud as it wends its way from an obscure back road to the field of combat on a breezy hillside. This is four-wheel-drive country. No Nissan Micras here.

Welcome to the weird, wonderful, windswept world of vintage ploughing, the pastime which is to motor sport what slow bicycle racing is to the Tour de France. Where the tea comes in half-gallon mugs and the bread in the sandwiches has the impossible freshness that only eating out of doors can impart.

The manufacturers who developed the rugged little machines which mechanised and modernised agriculture in the forties, fifties and sixties have a great deal to answer for. The tractors churned out of their factories simply refused to lie down. They are still on the go – and appearing at a vintage ploughing match somewhere near you.

So what have we here today? Ford and Fordson Dexta, Chalmers and Allis Chalmers, something called a Cropmaster, David Brown of course, McCormick, Case, Ransome and an Ursus, its lines as sleek as an explosion in a Meccano factory. There is even a Porsche, though the resemblance to the sports car of the same name ends after checking that both have four wheels. There are no marks awarded here for looking smart. This is not a gymkhana.

Then there are the Massey-Fergusons, like a shoal of eager grey fishes. One ingenious owner has clamped a makeshift cab on top of his Massey. It resembles the helmet worn by outlaw Ned Kelly. But no one dares to tamper with an engine that still purrs steadily beneath the bonnet after four, maybe five, decades of heaving and hauling, fetching and ferrying. Admirable. Indestructible.

For those who take the word 'vintage' seriously, there are the diehards who insist on bringing, not a tractor, but a pair of piebald steeds. Ploughing with live animals, otherwise the preserve of farmers in the Third World, continues to hog the attention of the photographers at such events. Privately, even those who retain the horse in 'horsepower' admit that the beasts eat far more than they are worth.

The ploughing gear is complicated. Rigs first assembled over half a century ago still look like works in progress – all weights and welds, bolts and bars. Everything is customised and adapted to the whims and convictions of the owner. And at the cutting edge of each Heath Robinson assemblage of jumbled metalwork is a blade of cold, clean steel. Look at the trailers, for instance: those rickety relics of the days before hydraulics came as standard. The cutting plate shines out from under the struts and chains and discs.

The rules of competitive ploughing are obscure, with more local variations than a jazz combo reworking an old standard. Each contestant is given a small plot of stubble field to plough – a task that takes several hours. If real farming took place at this pace, with such painstaking attention to detail, there would be no such thing as commercial agriculture.

The sport has a language all of its own. Adherents speak of 'scribing' and 'the split'. They make constant adjustments using a spanner, a lump hammer, a measuring tape and a set of sticks. Their goal is to lay the brown earth bare in a manner that imposes a mathematical formula of millimetre-exactness on a piece of stubble field.

The satisfaction to be gained from performing the task elegantly and accurately reflects the imposition of human will on the wilds of nature, which is the very foundation of civilisation.

6

OLD BOY RACERS

Vintage fans are the boy racers of the tractor world. The idea behind preserving the old machines is to have something stylish to show off in front of your mates, without actually busting the bank.

In his parallel milieu, the boy racer plays with, and displays, a clapped-out but stylish Toyota Celica. The vintage-tractor devotee gives his heart to a superannuated Case or Field Marshall which looks dilapidated but still packs a few horsepower beneath its bockety bonnet. The fact that the up-to-date modern tractor is faster, bigger, cleaner and altogether more efficient than such machines is neither here nor there, as far as he (or, occasionally, she) is concerned.

Most of them are boys no longer, of course, and many of them have spouses who must occasionally harbour treasonous suspicions that the €1,800 spent on an old Ford from the fifties or sixties could be better deployed on a vacation or an appliance of science for the utility room. But in fairness, it must be accepted that a vintage tractor gives longer-lasting pleasure than any sunshine holiday or washing machine.

The real expense is actually in the little extras, like the shed as big as an aircraft hangar needed to accommodate the old jalopies and all the attendant paraphernalia. Then there is the trailer or low loader required to haul the ancient steed to all ploughing matches within a 150 mile radius on winter weekends.

The fixation with such hardy equipment springs in part from a nostalgia for the days when man and machine had to work in closer partnership than they do today. A high-tech, reversible five-furrow mastodon is probably capable of turning over an acre in not much more than half an hour. No wonder so many real farmers are part-timers. Those big beasts even come with high-powered lights so that the farmer can plough at night after putting in their shift on the building site or behind the counter at the bank.

The vintage enthusiast is in no such hurry. The vintage enthusiast is happy to experiment with an engine that requires more coaxing to start than an ice maiden faced with an unwelcome suitor. The vintage enthusiast is prepared to tinker, to take apart, to reassemble, to go to endless lengths to find spare parts in far-off garages or in the most obscure corners of the internet.

Some vintage enthusiasts are even ready to dabble with a fuel called TVO, which requires the tractor operator to switch from petrol to oil at just the right time. Doing so successfully is as delicate and rewarding a process as cooking the perfect soufflé, a miracle of precise temperatures and timing.

Then there is the oiling of the mouldboard – that's the plate of steel responsible for turning over the sod. A well-cared-for mouldboard can be just as good and effective as it was when made a century before. Achieving such burnished perfection demands the sort of dedication and persistence the owner of a Crufts-winning Afghan hound must lavish on the impossibly hairy dog's coat.

Though much of the machinery of yesteryear has perished, a great deal still remains to be evacuated from forgotten sheds along forgotten lanes. The Ferguson Black, for instance, is rare beyond price: it was a one-off prototype which is now kept prominently displayed at the Science Museum in London. Owning a Ferguson Brown, on the other hand, is a perfectly reasonable ambition for a vintage enthusiast who has moderately deep pockets. The Ferguson-Browns were mass-produced by Harry Ferguson and David Brown before World War Two and were so well made that hundreds of them have survived into the new millennium.

And so it goes on. Every make and model, from the steam-engined, iron-wheeled hulks of the early twentieth century through to the Fords and Fiats of more recent vintage – they all have their fans.

Nostalgia nuts – or boy racers at heart.

7

PLOUGH MUSIC

Composer Dmitri Shostakovich was born in 1906, a long way from the muck of any farmyard. The son of an engineer, he was very much a child of the city. The city in question was called St Petersburg when he arrived at 2 Podolskaya Ulitsa, and had become Petrograd by the time he enrolled at the local *conservatoire*. No mere school of music was good enough for our Dmitri, you note: it had to be a full-blown *conservatoire* for him.

Now, the reason Shostakovich merits a mention here is that, along with the fifteen symphonies, seven concertos, at least seventeen string quartets, seven operas, four ballets and scores for thirty-six films (including a couple of animated movies), the prolific Russian also dashed off a polka called 'Dance of the Milkmaid and the Tractor Driver'. Aha!

He fitted it in as part of his *Limpid Stream* suite, which was composed between the Suite for Jazz Orchestra No. 1 and the Sonata in D Minor for Cello and Piano during a period of raging artistic hyperactivity in 1934. He scored 'Dance of the Milkmaid and the Tractor Driver' for two violins – and a lovely piece it is too, in the correct hands, at once a genuine dance and a reminder that Dmitri was a musician of formidable technique who demands real skill of anyone wishing to do justice to any of his work.

Now, the question is why, if Shostakovich could write a pleasing polka about a tractor driver, the horny-handed sons of the harrow have been so singularly overlooked by other leaders of modern music? Dmitri was more a boyo for the drawing room than the machinery shed, yet he found it possible to cull inspiration from the challenge of one ploughman's epic struggle with nature and how he might come home from a day's labour ready to flirt with a doxy from the dairy. The ability to identify with the lot of the foot soldier in the perennial battle to produce sustenance stands to the credit of the man from the city that is once again called St Petersburg.

Yet mysteriously, ploughing has not been a fertile hunting ground for other songsmiths or composers seeking starting points for their hits. The only other song with a similar title that springs to mind is the number from *Oklahoma* which suggests that 'the farmer and the cowboy should be friends' – words by Oscar Hammerstein, tune by Richard Rodgers. The song is the warm-up for a dance-hall brawl – and a long way indeed from Dmitri's *conservatoire*.

The Beatles had hits with songs about paperback writers, tax officials and chauffeurs – no sign in their output of any polka-prancing tractor drivers. Unless, perhaps, just possibly, the poor addled 'Fool on the Hill' was such a one bereft of his milkmaid. Probably not.

The Rolling Stones, insofar as they touch on any food-production themes, prefer poultry to arable, in the form of 'Little Red Rooster'. Abba opted for the battlefield – 'Waterloo' – rather than the ploughed field. And so on.

The time was, *fadó fadó*, when the merry ploughboy was the main man in many a ballad, but that time has long gone. Though Elvis Costello may address issues such as Alzheimer's disease in his songs, and Billy Joel deals with post-traumatic stress disorder among US Vietnam veterans, the commonplace drama of agriculture is largely overlooked in pop.

You have to go back to the Victorians – and even further back – to find our particular pet subject given the full oompah-pah. Turn to No. 308 in the Church of Ireland hymnal:

> We plough the fields and scatter
> The good seed on the land,
> But it is fed and watered
> By God's almighty hand.

The hymn was written in German by Matthias Claudius (1740–1815) and translated into English by Jane M. Campbell (1817–78)

8

DEERE JOHN

The John Deere advert claims that 'More than 170 years' experience is reflected in every product'. The ad, by the way, is for lawnmowers designed to gussy up the gorgeous gardens of the aspiring middle classes.

What the original John Deere (1804–86) would have made of such frivolity is anyone's guess. Suffice it to mention that he was reared in a school of hard knocks, his chances of a more formal education scuppered by the fact that his father was lost at sea when he was just four years old. No doubt somewhere in the workings of today's 'EZtrak zero turn mower' is some spiritual residue of John's seminal steel plough of 1837. Somewhere.

A native of Vermont in sweet New England, he came up through the ranks of a blacksmith's apprenticeship before flirting with poverty and bankruptcy for much of his early adult years. His blacksmithing expertise in coachwork was not in great demand during the economic turbulence of the Panic of 1837, so he switched tack and hit pay dirt with the plough that cut such a clean, swift furrow through the glutinous soils of the Midwest as he moved to Illinois.

His ingenuity spawned a global empire that now employs over fifty thousand people. Compare that mega-outfit with the enterprise of 1849, which had a staff of around sixteen and a modest annual output of just 1,136 ploughs. More than a hundred and seventy years on, production may be scattered across Europe, Africa and the Americas, North and South, but the headquarters remain in Illinois, at Moline in Rock Island County, where the firm has been located since 1848.

Deere himself stepped down from command ten years later to concentrate on local politics, as mayor of Moline, and on the breeding of Berkshire pigs. He was happy to leave his son Charles in charge of the Moline Plow Manufactory, as it was grandly styled.

The Deere family's direct influence on the company expired before a century had elapsed, but the name lives on, and how! There is an argument for suggesting that some of the glory should have gone to Gilpin Moore, for it was Moore who came up with the 'Gilpin Sulky Plow', which vanquished all rivals at the Paris Universal Exposition of 1878. The 'Sulky' allowed the ploughman the unheard-of luxury of sitting while at work, but its inventor, though inspired as a designer, was no more than a hired hand of the Deeres.

Similarly, it is plausible to argue that the Deere brand could just as easily have been the Waterloo Boy marque after the company lurched reluctantly into making tractors in 1918. Their own early efforts progressed little further than a weird prototype with a ferociously complicated arrangement of clanking chains. Instead of heading further down that particular cul-de-sac, they acquired Waterloo Boy, brainchild of internal-combustion-engine pioneer John Froelich.

Let's be honest here, Froelich would never have captured the imagination of American songwriters in the same way as good old reliable John Deere. (What the hell rhymes with 'Froelich'?) Taking into consideration the firm's stubborn, prolonged insistence on sticking with two-valved engines until the 1960s, the name John Deere was tailor-made to become the iconic tamer of the prairies and friend of the hard-working farm family.

> I'm like a John Deere tractor
> In a half-acre field
> Tryin' to plow a furrow
> Where the soil is made of steel.
>
> THE JUDDS

Aside from EZtrak mowers, bunker rakes, barbecues, safes, wind farms, articulated dump trucks, excavators, marine engines and, yes, tractors, Deere & Company still produces ploughs – a hundred and seventy years, and counting.

9

THE BLESSED TRINITY

Psst! You interested in ploughing, yeah? Then what you need is 'The Book'. The Book is called *The Tractor Ploughing Manual*. And it comes all the way from Ipswich.

Of course, they know about ploughing in Ipswich. After all, flat and infinitely fertile East Anglia is England's answer to the Donetsk Basin, not to mention the historic headquarters of Sims, Jeffries and, most particularly, Ransome – plough-makers to the stars.

Thanks are due to John Walshe of Adamstown for making sure that this no-nonsense illustrated volume, edited by Brian Bell, was not overlooked in the current compilation. However, in making his case for the importance of The Book, John may have exaggerated ever so slightly the accessibility of the manual to the non-ploughing reader.

Reducing this subject to mere sentences and paragraphs makes for a text that is about as easy to follow as a knitting pattern for an Aran sweater. If you do not already know your plain from your purl, you may struggle to keep up. Nevertheless, let us see what we can learn from *The Tractor Ploughing Manual*, with its stream of references to bar point bodies, headstocks and heel irons.

For starters, Brian Bell — former vice-principal of Otley Agricultural College — attempts to put some measure of coherence on the assemblage of ironmongery that is the basic modern plough. Look more closely into any conventional tractor-mounted unit to distinguish the coulter, a metal disc with a sharp edge which makes the initial incision into the earth.

The coulter is followed in due course by the pointed share, which does the serious heavy-duty cutting of the sod. Then along comes the mouldboard, ready to lift and turn the earth, with all the slick ease of a short-order cook flipping an egg easy over. On these three components hang all the law and the profits: coulter, share and mouldboard. In that order.

The simple trinity of coulter, share and mouldboard is open to an infinite variety of refinement and a multiplicity of add-ons. The editor of The Book commends the slatted mouldboard, for instance, as being useful in tackling sticky soils — much more efficacious than a general-purpose or semi-digger mouldboard. He reminds us that, in terrain where large stones pose a hazard to machinery, it may be necessary to fit a spring to the share in order to reduce the risk of damage. When tackling grassland, a skimmer may be used to ensure that grass and weeds are properly buried.

Such neatness as can be achieved by judicious use of a skimmer is especially appreciated by those who enter competitions. Where ploughing for private gain is approached in a slam-bam-thank-you-ma'am frame of mind, those who cut a furrow as a hobby tend to be more appearance-conscious than the most neurotic of housewives waiting for the mother-in-law to call.

The book provides good advice on how to impress the old dragon (or, in this context, the judges). Adjudication is informed by a strict scoring system that looks for clearly defined, uniform furrows that are as straight as the side of any triangle found in Euclid. They have a tariff of penalties for those who fail to dig deep enough, leave unsightly wheel-tracks or finish facing north when they should be pointing south. (It's true!)

Those who expose themselves to the verdict of such nit-pickers would do well to fine-tune the gap between their mouldboards. They would also do well to find a flat, firm surface on which to check that their shares are in line. They will then enter the fray confident that their shares are at the correct pitch (angle to the ground).

This is care and precision of the sort demanded on an Apollo space mission, applied to an open field rather than outer space.

10

THE GENTLEMAN'S TRACTOR

It was back in 1938 that the Minneapolis Moline UDLX – nicknamed the Gentleman's Tractor – hit the market with a resounding thud.

The fully glazed cab of the UDLX offered such seductive delights as a radio and a heater. Specifications included windscreen wipers, a cigarette lighter, a glove compartment and a foldaway passenger seat for the gentleman's lady. The idea was that the vehicle which was used for ploughing all day could then double up as a 6,400-pound runabout when it was not required for agricultural duty. Its 39 horsepower engine could be coaxed to hit a thunderous 40-plus miles per hour on the open road. Minneapolis Moline drummed up a crowd of twelve thousand farmers – or so the legend goes – for the launch of the

Gentleman's Tractor. Unfortunately for them, the allure of all those added extras, such as the speedometer and the window defroster and the rear-view mirror persuaded only a hundred and fifty customers to part with the $2,150 the firm was asking for it.

Some concerned reviewers suggest that the brakes fitted to the UDLX were not quite up to the job of stopping this monster safely when it was clapping along with full formidable momentum. Perhaps that was a factor in its commercial failure. Or maybe potential customers, still suffering the aftershocks of the Depression, were prepared to risk sunburn and settle for a smaller, less ostentatious Ford 9N, which came on the market in 1939 with a price tag of $585 (radio not included).

Of course, we are all gentlefolk these days. We all love our gadgets. And we are not averse to the occasional gimmick. No one makes tractors now without cabs – if only for safety reasons. The time has gone when some Massey-Ferguson models boasted 'drapes' designed to funnel some of the incidental heat generated by the engine back towards the shivering driver – if the wind was blowing from the correct quarter.

Today, much farm work is conducted from within a sealed, computerised, air-conditioned bubble. The noise level inside the bubble is kept at a very respectable 70 decibels, thanks to the bulkhead that maintains the occupant, in ergonomic splendour, at a safe remove from the roar of the engine. The bubble, with its tinted glazing, has its own suspension independent of the rest of the tractor – no doubt in an effort to keep spillages from the drinks holder to a minimum. For sheer animal comfort, the driver of an up-to-date tractor has the edge over any mere car driver.

Climb up the steps into the cab of the latest Deutz Fahr tractor and take a seat. There are twenty-three knobs, dials and joysticks available at your right hand, never mind the standard driving instruments straight ahead, plus a few more switches and controls at the left. The soloist playing Bach's 'Toccata and Fugue' on a full cathedral organ has just a few more things to fiddle with.

The combination of controls on the Deutz – a brand typical of the modern breed – make twenty-four different gear speeds available, covering the range from 3.45 to 50 kilometres per hour. Only a potato farmer could ask for more – so the manufacturers thoughtfully provide sixteen additional 'creeper' speeds, allowing the machine to operate at a snail's pace of 0.48 kilometres per hour.

The scope for in-bubble entertainment is immense. The radio is not even the half of it. At 70 decibels of ambient noise, there is no problem listening to a favourite CD. While the official dealers do not offer such an option, a fashion has emerged for installing a 12-inch television where the rear-view mirror should be. It would surely be possible to catch up with *Emmerdale Farm* or take an Open University course while chomping through thirty hectares of ploughing in a day's work.

Data are for the Deutz-Fahr Agrotron X710 or X720

11

Minnows – and Sharks

What follows is complicated, convoluted and potentially confusing. Readers with poor concentration levels are advised not to linger here. Rather, they should move on directly to other pages in search of more readily palatable entertainment. The few of us left will furrow our brows and knuckle down to a brief examination of how businesses which often started in the coals of some humble blacksmith's furnace were sucked up in the grand sweep of industrial globalisation.

Brands of tractor often had their origins in the obsessions and enthusiasms of individual inventors. This is a summary of how such minnows were devoured by bigger fish, which in due course fell into the maws of great sharks. Some inventors left their names on bonnets, while others were lost in the whirlpools of mass production, where the shots are called by transnational boards.

AGCO is one of the biggest of the big sharks. It has American roots, and the 'A' in the name provides a link with the Allis heritage – a tenuous tribute to Edward Allis, who died in 1889, long before the Model 10-18 that bore his name went into production, in late 1914. The 'G' stands for Gleaner, a line of combine harvesters, while the 'CO' is presumably for Company.

They have an impressive portfolio among their world-wide interests. Massey-Ferguson (Canada), for instance, was something of a collector in its own right. Ferguson, the Ulster-born tractor tycoon, was associated with David Brown (UK) and then Ford (USA) before he finally merged his name with Massey Harris (Canada) to form Massey-Ferguson in 1953. Massey-Harris had already pulled in Wallis (USA) in 1928.

Massey picked up Landini (Italy) in 1960, though it was later sold off. They also had an interest in Ursus after the Polish company had dallied first with Lanz (Germany) and then with Zetor (Czech). But Massey-Ferguson (Canada) was no longer its own master after AGCO stepped in, around 1992.

And the aggrandisement did not stop there. McConnell, which made large four-wheel-drive tractors along Massey-inspired lines, was acquired by AGCO in 1994. Iseki (Japan) was linked to White (USA) for a while before becoming part of the growing AGCO empire in the early 1990s, along with White itself. Fendt (Germany) was picked up by AGCO in 1997.

White came to the party with its own considerable bundle of past annexations. Hart-Parr (USA) merged with Oliver (USA) in 1929, and it was the Oliver name that proved the stronger brand. Cletrac (USA) came under Oliver (USA) in 1944. But the gobbler was gobbled in 1960, when White (USA) had the deep pockets. The last tractor carrying the Oliver badge was made in Iowa in 1974.

Cockshutt (Canada) lost its independence to White (USA) in 1962. Minneapolis-Moline (USA) was picked up some time around 1963, and the new owners dropped the Minneapolis name in 1974. Massey, Ferguson, Harris, Wallis, Cletrac, McConnell, Iseki, White, Fendt, Cockshutt, Minneapolis, Moline, Hart-Parr, Oliver – quite a compilation.

Here is another case (ahem) in point. The Case in question was Jerome Increase, or 'J.I.', of that name, a Wisconsin senator and several-times mayor of Racine. Jerome Case died in 1891 – another man who was well out of the picture before the business he founded really began to take off.

The acquisitory streak first began to manifest itself in 1928, when Emerson Birmingham (USA) became part of the JI Case corporation. Rock Island (USA), which began selling Heider tractors in 1914, was taken over by Case in 1931. SFr/Vierzon was enveloped by Case in the late 1950s, and the name of the French concern disappeared a few years afterwards.

Case had to be bailed out by Tenneco (USA) in 1967,

when Wisconsin tradition was rescued by Texas capital. Once they had their toe in the water, the Texans continued to splash out. David Brown (UK) was taken over by Tenneco in 1972. More significantly, International Harvester (USA) followed into the same stable in 1984. Steiger (USA) was snapped up by Tenneco (USA) in 1986 and added to its Case compendium. In 1999, Case-IH, as it was then called, merged with New Holland (see below) – to the alarm of governments, concerned at the possible effect on competition of such a massive agglomeration.

After a brief foray into the tractor market, General Motors skedaddled to pursue other interests after 1922. However, given the farming background of founder Henry Ford, it was always likely that GM's great American rivals would be in for a longer haul – seventy-four years from 1917, as it turned out. The Model F was the first in a string of Fordsons and Fords that helped to mechanise agriculture on the grand scale.

In its later years of involvement in the business, Ford was not averse to growing through purchase, though this proved to be no more than a defiant departing thrash. New Holland (USA), previously best known for its farm implements, made its first tractor as late as 1986, after it had been bought by Ford (USA). The fresh group also absorbed Versatile (Canada) around the same time.

However, Ford then sold out their agricultural divisions to Fiat (Italy) in the early 1990s. This already vast business later merged with Case IH in 1999 to produce the gigantic CNH International. Somewhere in the mix was the French concern Someca. To recap, the Fiat snowball has rolled up Ford, New Holland and Case (with all that is therein entailed), not to mention Someca. The Case spectrum extended across Emerson Birmingham, Rock Island, Heider, Vierzon, David Brown, International Harvester and Steiger.

John Deere (USA) is another major presence on this scene, but the folk in Illinois have never dabbled as extensively as some of their adversaries in the used-tractor market. Nevertheless, it was their purchase of Waterloo Boy (USA) in 1918 that first moved them into serious tractor manufacture. John Deere also bought the Lanz factory in Germany in 1956, and Chamberlain (Australia) fell to Deere in or around 1970.

The predatory instinct is more keenly exhibited in SAME of Italy (the initials stand for Societa Anonima Motori Endotermici). Guldner (Germany) was absorbed by Fahr (Germany) in 1962. Deutz (also Germany) merged with Fahr in 1968. Then along came SAME in 1995, when they acquired the combination for $65 million. Hurlimann (Switzerland) had been owned by the same SAME from 1975. The Italian buyers had already put their compatriots Lamborghini in their large shopping basket during the 1960s.

Deutz was delivered with some neat US genes in its make-up. The long-gone Advance Rumely (USA) was a company with a perfectly decent pedigree: it made the transition unaided from steam threshers to lumbering farm tractors in 1910. It then remained proudly in the field until it was finally taken over by Allis Chalmers (USA) in 1931 – yes, named after the same Edward Allis who supplied the 'A' in AGCO. Don't ask. Allis Chalmers was eventually subsumed into Deutz in 1985. Deutz, in turn, was absorbed by Italian multinational SAME.

Other European operators could not match SAME or Fiat for scale, though they did not lack the ambition. Porsche (Germany) in 1957, and then Remault (France), which made its first tractor as far back as 1918, took a proprietorial interest in 1964.

In Scandinavia, Bolinder-Munktell (Sweden) merged with Volvo (Sweden) to become Volvo BM in 1950. Somewhere along the way, they also amalgamated with the Finnish state-owned Valmet. The Finns bought out the Swedes in 1985 and the operation was privatised as Partek in 1997, promoting the Valtra brand.

All of this intricate material comes from close examination of the splendid *Ultimate Guide to Tractors* by Jim Glastonbury, Grange Books, 2007, which is exactly what it claims to be.

12

THE 1994 NATIONAL PLOUGHING CHAMPIONSHIPS

Only the Tour de France (in 1998) and Fleadh Cheoil na hÉireann (in 1999 and 2000) have put modern Enniscorthy as glaringly in the spotlight as the National Ploughing Championships held in the County Wexford town in 1994. For three days, the farm of Charles and Jennifer Kavanagh in Drumgoold, a short walk up the Shannon from the centre of the town, became the Mecca for 150,000 or so agricultural pilgrims.

It turned out, of course, that the ploughing was actually a very small, peripheral part of the excitement at the championships, though there was a respectable-sized crowd keeping an eye on the efforts of Austrian competitor Helga Wielander. She was a veritable vision, with a spanner in her hands and fetching blue wellingtons on her feet.

The horse-ploughing classes also attracted a sprinkling of onlookers – an audience who approached the quaint activity with more bluntness than sentiment. 'He's tearing the arse out of it,' declared one jaundiced spectator before returning to the buzz in the main exhibition field.

What had been a meadow a few weeks earlier now resembled the battle headquarters for some mad medieval army – so many tents bedecked with flapping flags. Every bank, every machinery manufacturer, every distributor of agrichemicals was represented somewhere in this madhouse bazaar. The Waterford Foods balloon was the highest flier among the swarm of dirigibles that floated overhead.

There was room too for the balloon sellers, the burger vans and the gizmo hucksters who thrive wherever there is a crowd. One of the latter was flogging the latest word in vegetable and fruit peelers, letting likely customers have a go before purchase. The result was a mountain of potato flitters and a pile of butchered spuds.

The band, under the baton of Father Willie Howell, played 'When the Saints Go Marching In'. The smell of pine resin filled the air around the Cóillte forestry stand. Driver John Murphy waited for his boss, Minister of State (and local TD) John Browne, to launch a fluke and worm drench.

Clokes of Cloheden took bookings for their mobile

stages, and the Girl Guides garnered a few bob for the organisation by selling programmes. John Bruton, who has since become EU ambassador to the United States, appeared, glad-handing the crowd in Charlie Kavanagh's bar. The pub had been miraculously transposed, lock, stock and beer barrel, from the bottom to the top of the Shannon Hill for the duration. The counter was approximately a mile long, and Maura Flannery sang the blues for four hours straight, running on nothing more intoxicating than iced mineral water.

Press photographer Paddy Murphy, on the prowl for the Enniscorthy paper, was in trouble as he attempted to pick out the Wexford folk from among the throng. Everyone, but everyone, he stopped was from Askeaton in County Limerick.

John 'Muzzie' McCrea was triumphant at the way that he and his rugby-club pals managed to keep things under control in the car park assigned to their stewardship: 'They are not just parked, they are colour-coordinated,' he boasted as he reached for a well-deserved Guinness. Indeed, he was entitled to his boast, for the business of stewarding can be every bit as complicated as the furrowing, opening, skimming and packing of the ploughing competitions themselves.

Over the road from the main field, on Paddy and Brenda Kavanagh's land, there were rare breeds and quare breeds, for sure, among the cattle on display. The Aberdeen Angus were the obvious draw: black, beautiful, beefy – and a real throwback to the succulent steaks of yesteryear. On the other hand, perhaps the Piedmontese, or even the Montbeliarde, are the livestock of a future set to be dominated by Continental strains. Or what about the Meuse Rhine Issel, a dual-purpose breed with a triple-barrelled name?

Then there were the ostriches: 'If that yoke gave you a belt of a kick, it would be worse than a kick of a cow.' Thank heavens the long-legged African beauties were well corralled, so.

> In 1960, reigning 'Queen of the Plough' Mary Shanahan, from Kerry, was to the fore at the Saint Patrick's Day parade in Dublin. Very fetching she looked too, as she piloted her tractor around College Green in front of massed crowds. Whether they were admiring her tiara or her Pierce-made two-sod plough is impossible to tell from the photos that captured the event. The initial 'Queen of the Plough' was another Kerrywoman, Anna Mai Donegan, who took the treasured tiara in 1955 and again in 1956. Three-in-a-row titles were secured by Marion Stanley of Cork (1985–87) and Michelle Kehoe of Wexford (2000–02, there being no event in 2001).

Adapted from the *Guardian* of 29 September 1994

13

THE PSYCHOLOGY OF PLOUGHING

Ploughing has nothing whatsoever to do with sex. Small wonder that it is an activity completely ignored by the rappers and rockers of pop culture, with their discos and base, city-centred obsessions. It is, more often than not, carried out during the colder months, the shorter days, the worst of the year's weather. It is hard to counter the suggestion that ploughing is a solitary activity and really rather, ahem, boring.

The sowing and reaping and (goodness help us) threshing which follow ploughing all reek of reproduction, or at the very least flagellation, it cannot be denied. In contrast, the initial ritual turning over of good brown earth is more a wiping clean of the slate – more an absolution than an act of seduction, procreation or titillation.

The making of the bed holds the casual observer less enthralled than the dirtying of the linen – just ask any newspaper editor or moviemaker. Ploughing is worthy and necessary but it is also repetitive and sterilising. At the end of the process, we have harvest festivals, happy communal celebrations. At the start, we have solitary, plodding coverage of vast acreages.

Yet this apparently lonesome occupation is something that many landholders confess is their favourite part of farming. They look forward to it and prepare carefully for it. The simplicity of purpose and execution may help to explain its enduring fascination. The chore is underpinned by generations of experience, with no chemicals involved. It is also the ultimate marking of territory by the human species. Where other jobs may be hived off to contractors, the ploughing is generally the last piece of the arable process that is allowed to slip into the control of outsiders.

On a drab February night, beside the steady thrum of traffic along a national road, headlamps may be seen buzzing around a field as two massive tractors are at work. Ploughing. In the dodgy, starless light, they resemble two mechanised gladiators in a fantasy film. The lads in the cabs confess that they are in a hurry. This is not a competition. There are no marks for splits, straight lines or tidy ins-and-outs. This is no more than a job, and one that has to be completed quickly.

The horizons of these men are not limited by the

ditches that surround this fine big field. The chances are that they have alternative jobs away from agriculture in offices or on building sites. Yet they still insist on coming back to base to run their ploughs over the family holding. People who come from arable land simply hate to see that land idle. Like so many squirrels storing up nuts, they are driven by instinct to repeat the actions of previous generations.

Those previous generations marked out the boundaries of the fields.

Those previous generations doubtless spent – and resented – many hours of childhood picking stones to save wear and tear on equipment. Perhaps the usually solitary nature of the task helps to put the ploughman in touch with those previous generations. It is also, of course, an opportunity to play that Metallica CD the girlfriend refuses to allow anywhere near the stereo.

Besides, perhaps it is not so solitary a process at all. The smell of fresh soil brings the inevitable swarms of rooks, picking along the furrows in search of whatever morsels have been thrown up. There are so many of them, indeed, that the tractor leaves the field with its paintwork covered in a white film of bird droppings.

Look closer at the newly turned earth to catch a glimpse of a fat worm, its body gleaming in among the muck as it dives to evade the greedy rook. Not so solitary after all.

With thanks to the closet Abba fan

14

MAKING PROGRESS

British farmer and broadcaster Alfred Hall is a one-man fount of lore about all things agricultural. Our Alf once put forward a very credible, authoritative (and quaintly sentimental) explanation for the emergence of ploughing as a sport. The Cumbrian countryman traced the competitive element in what is otherwise straightforward good agricultural practice back to 'moving day'.

The phrase has more recently been adopted by professional golfers and applied to the third of four rounds in a tournament as the competitors jostle to put themselves in contention for the big money. To Alfred Hall MBE, however, moving day, or Candlemas, or 2 February, was significant as a landmark on the calendar among the tenant farmers of Scotland and northern England. Over centuries, this was the date when tenancies came to an end, prompting a migration of many farm families in an agricultural version of musical chairs as they transferred from one holding to the next.

The net result of all the moving was that new occupants would find themselves under stress and harassed, with all their worldly goods piled up higgledy-piggledy on a cart, in the yard of their latest home, on 2 February. Spring would be stirring, and the newcomers were faced with an immediate, frantic backlog of urgent work if they were to make a success of their fresh assignment. The most pressing and daunting task of all to be tackled was preparing the fields for planting cereals.

Fortunately, as Alfred explained in his book *Ploughman's Progress*, the tenant farmers of the border country were not professional golfers. No, they were genuine team players. Padraig Harrington or Eldrick 'Tiger' Woods, for example, may be the nicest guys in the world, but their moving days are strictly first person singular. Not so these country folk, who thought nothing of rallying around those in need come Candlemas. Up to a dozen of them would all just happen to drop by, with their horses and ploughs, on the same day, all offering their services for no more than a cup of tea and a slice of bread.

Such good neighbourliness served as an opportunity for one ploughman to improve his technique by watching other ploughmen at work. Good neighbourliness was an opportunity to compare equipment and to discuss such matters as the ideal depth of a furrow. Good neighbourliness also bred competition and, this being Britain, it was only a matter of time before someone came up with rules and regulations. The island that codified soccer, rugby, cricket, hockey and the rest had no problem coming up with a system of marks for ploughing.

Alfred Hall's heart-warming tale of decent folk rallying to the assistance of fellow farmers has the ring of truth to it – but maybe not the full truth. When Alfred and his friends in Cumbria attempted to stage a ploughing match in 1947, they invited four foreign ploughmen to join them in the field at Calva Farm in Seaton. The visitors were in Europe as the winners of the 'Ontario Championship Plowing Match'. You may bet your bottom (Canadian) dollar that these boys knew nothing about Candlemas or moving day on their windswept prairies.

The notion that ploughing competitions were somehow the preserve of those who came from the shadow of Hadrian's Wall is strained by consideration of what happened next. The match at Seaton was called off due to freezing conditions. However, the germ of an idea had been sown, and an international fixture was finally staged in early 1948, with Canadians joining Scottish, English and Welsh participants.

By 1953, there was enough interest around the globe to have a world championships event in Ontario; it was won by local entrant Jim Eccles. Within ten years, Australia, Austria, Belgium, Denmark, Finland, France, West Germany (East Germany was not involved until 1965), Italy, Netherlands, New Zealand, Northern Ireland, the

PLOUGH MUSIC

Republic of Ireland (which staged the second championships, at Killarney), Norway, Sweden, the United States, Yugoslavia and Zimbabwe (then Rhodesia) were all on board.

Candlemas is doubtless big in Zimbabwe, Alfred.

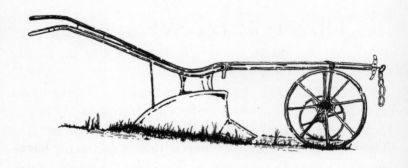

Ploughman's Progress by Alfred Hall (Farming Press Books, 1992)

15

TRACTOR HEAVEN, USA

'From the town of Lincoln, Nebraska'
     BRUCE SPRINGSTEEN

The Midwestern state of Nebraska must be tractor heaven, with the state university in the city of Lincoln at the centre of this bucolic nirvana. The university's Professor Roger M. Hoy sums it up: 'Nebraska leads the US in beef production, which occurs primarily in the West, and is a major contributor of corn, soybeans and other row crops in the East.' When Americans speak of 'country', it is probably Nebraska they have in mind.

Most folk in Ireland reckon themselves no more than a generation away from the dung heap and the hay shed. Yet the population of the island is packed into the thirty-two counties at a density of around 183 persons per square mile. At this stage, the Irish are more typically apartment dwellers than farmsteaders.

Nebraskans, in contrast, are spread considerably thinner across their ninety-three sparsely peopled counties. With just twenty-two of them rattling around the average square mile, they are the real rural McCoy. Thus it is that one of the main draws at the Saunders County Fair in Wahoo (population 3,942 – on a par with Listowel) each year is the arrival of the Nebraska Bush Pullers – thanks to the blessed sponsorship of Sapp Brothers Petroleum.

The bush pullers are enormous tractors with more hulking horsepower than sense. They appeal to the fantastic in the imagination of every farmhand who has ever sat in the cab of a homely John Deere. This is a land where tractors, and ploughing, matter.

As the denizens of Nebraska (including Professor Hoy) will never tire of telling you, it was back in 1919 that the state first passed legislation requiring tractor manufacturers to submit their machines for rigorous testing by the boffins in the agricultural-engineering department of the university. As a result of this law, they have been running the rule over the various makes and models since 1920. The professor is the current director of the Nebraska Tractor Test Lab, where they have an unmatched reputation for probity and unbending adherence to their own high standards. Yes, there are other testing centres in the likes of China, France, Germany, Italy, Japan and Spain, but none of them carries quite the same clout as Nebraska.

The results tend to be delivered more with figures than with words. The reports are not long on humour — at least not unless you look hard . . .

The Bates Steel Mule 15-22 Model F, made by the Bates Machine and Tractor Company of Joliet, Illinois, was tested in October 1920. The lab team concluded: 'In the advertising literature submitted with the application for test of this tractor, we find some claims and statements which cannot be directly compared with the results of this test as reported above. It is our opinion that none of these are unreasonable or excessive, with the exception of the following:

' "The Bates Mule is the most efficient tractor in America, barring none"; "The Bates Steel Mule will always work equally well in wet or dry soil, good or bad conditions"; "The Bates Steel Mule is a perfect field machine".'

The brewers of Carlsberg (probably) recognise what Bates was going through.

The first mention of Ferguson at East Campus was in 1948, when the Ferguson TE-20, manufactured in Coventry, fitted with Girling brakes and powered by a Continental engine, was put through its paces by Lester F. Larsen, the renowned engineer in charge, and his team – in test 392.

In the space allowed for remarks, they contented themselves with observing that: 'During test "B" a decrease in horsepower occurred. The head was removed and the combustion chamber cleaned; improved horsepower performance resulted. Fan belt was replaced with a new one during preliminary belt run.'

Under Professor Hoy, such highly embroidered and evocative commentary would be frowned upon as a case of the runs. Here is the summary of test 1892, carried out on a John Deere 5403 Diesel 9-speed in the summer of 2007: 'All test results were determined from observed data obtained in accordance with official OECD, SAE and Nebraska test procedures.'

So now you know.

> In 1905, a gentleman named C. M. Eason collaborated in Kansas with Ansel Wyong to make a prototype tractor that had just one rear wheel. The single wheel at the back proved problematical, so the pair then experimented with two rear wheels and two in front. However, the steering wheels to the front were set so close together that it was, in effect, a tricycle tractor. Eason later joined the Wallis Tractor Company.

For tractor test reports, try: *http://digitalcommons.unl.edu/*
For the population of Nebraska:
http://quickfacts.census.gov/qfd/states/31000.html
Thanks also to Professor Roger M. Hoy

16

THE CRADLE OF THE WORLD PLOUGHING CHAMPIONSHIPS? ER, WORKINGTON . . .

Workington is a port town on the west coast of Cumbria, in the north-west of England. Workington, ancient market and industrial centre at the mouth of the River Derwent. Workington, home of speedway's Comets and of soccer's Workington Reds.

Ah yes, Workington, where the harbour handles about three hundred ship movements annually. Workington, with a hundred-and-twenty-five-year history of rail-making at the furnaces of the Workington Iron and Steel

Company. Workington, you're famous, but really not famous enough.

Let's hear it instead for Workington, the cradle of the World Ploughing Championships. Yes, here in Workington is an example of a global phenomenon that began as a local initiative, taken by the Workington & District Agricultural Society.

The society was founded by progressive landlord John Christian Curwen (1756-1825), member of parliament for Carlisle and general good egg. Several generations on from Curwen, it was the members of the society who started the international ball rolling during the 1940s.

In an inspired attempt to provide some excitement and lift the post-World War Two doldrums, the committee made contact with a party of Canadian ploughmen who were touring England. The visitors were invited to take part in a ploughing match staged at a place called Stainburn Hall early in 1948. The event generated great interest and proved to be the start of something that grew to be phenomenally big.

Six decades later, the 2008 World Ploughing Championships were presented amidst much razzmatazz and corporate schmoozing at Grafenegg in Austria. Representatives of Northern Ireland, Austria, Finland, Wales, Scotland, the Republic of Ireland, France, Norway, England, Norway, Lithuania, Australia, the Czech Republic, Spain, the Netherlands, Sweden, Italy, Germany, Slovenia, Slovakia, New Zealand, Canada, Belgium, the United States of America, Denmark, Croatia, Kenya,

Switzerland and Hungary took part.

All the sport and technology and hardware, all the celebrations and the coming together of nations enjoyed in Grafenegg, followed on in a direct line of succession from the happenings at Stainburn Hall near Workington.

Yet the fact is that the pioneering events of 1948 have left little or no enduring mark on Workington, where it all started. The lack of interest in the event is confirmed by Pat Hall, manager of the town's Helena Thompson Museum, which concentrates on old costumes, crafts and furniture.

Pat Hall reports, with the air of a Methodist missionary cast among a bunch of staunch sun worshippers: 'I regret that at this time no special interest is shown in the agricultural history.' So, no parties of tipsy ploughboys wandering the streets of Workington late at night, and no queues of tourist buses waiting to gain admission to Stainburn Hall.

The strong tradition founded by the Right Honourable John Curwen has been allowed to wither. The last agricultural show presented by the Workington & District Agricultural Society took place in the 1960s, and Pat Hall, surveying the deserted altar of progressive farming, reckons there is no way that the once-annual event will be revived.

There is one small ray of hope. Pat is doing his best to spread the gospel in the teeth of the general indifference. A small corner of the Helena Thompson Museum has been reserved for matters ploughing. A special display, featuring a golden model of a plough, has been set up on the first floor as a reminder of the association between the mighty world ploughing organisation and the modest town where it had its roots in 1948.

Keep the faith.

The Workington 'Reds', by the way, play at Borough Park, and the team finished twelfth of the twenty-two teams contesting the Blue Square North league in the 2008-09 season, some 34 points adrift of winners Tamworth.

17

THE MAN IN THE MIDDLE

This section is dedicated, in fond sympathy, to Tom Henning Ovrebo. Tom, it will be recalled, is the Norwegian soccer referee who was hounded from Stamford Bridge by Chelsea players infuriated by his handling of their game against Barcelona in the 2009 semi-final of the Champions League. He was following in the unfortunate, haunted footsteps of fellow Scandinavian Anders Frisk, the Swedish official who handed in his whistle for good after an encounter with Chelsea that resulted in death threats being made against him.

This section is penned with Paddy Russell in mind, too. Paddy was the man in the middle at Fitzgerald Stadium when Gaelic footballer Paul Galvin decided to knock the ref's notebook from his hand. Attempts to pass the

Kerryman's petulant action off as a swat at a wasp were more laughable than plausible.

If Tom seeks deference, and if Paddy is keen to keep notes in his grip without interference on the field of combat in future, then perhaps they should both consider adjudicating at ploughing matches as an alternative career. The judges in such competitions never require police escorts to see them safely on to the road home after fixtures. They carry out their duties with no fear of molestation. Their ears are not troubled by shouts of unrestrained abuse from disgruntled protagonists, coaches or supporters. Though their verdicts are not always greeted with universal agreement, they are never openly challenged. It is simply not done.

Carlow native Tom Jenkinson is one such adjudicator. He has never been locked into the boot of his own car – or locked out of a changing room. There is scarcely a venue in the land where he has appeared to which he has not been invited back, and made welcome. He has presided at matches across the swathe of territory from Dunleer to Dunmanway without undue incident anywhere along the way.

It helps, of course, that ploughing judges generally operate in threes. There is strength in such numbers. On important occasions, they may even patrol in sixes. The points they award reflect the considered views of several experts measured against objective criteria, not whim or gut feeling. Challenging such august tribunals is patently a waste of time.

The ploughing judge is not required to bare his knees and turn out in black shorts. Nor does he don a bowler hat to draw attention to himself, as is the wont among the gymkhana set. Rather, he tends to blend into the background with modest attire, dirtying his old boots in among the rows. Michel Platini, please note.

Having spent fifteen years on the roads of Ireland, travelling to matches in Oylegate, Enniskerry and beyond, Tom Jenkinson reckons that Carnew in Wicklow is the best venue for atmosphere – a friendly place in the most friendly of sports. He suggests that the best catering is probably to be enjoyed at any of the matches staged in his home county of Carlow. The best land he ever saw for a ploughing competition was at Nolan's farm in Tullow, which hosted the 2006 national championships.

Like Ovrebo or Russell, Tom Jenkinson needs eyes in the back of his head, to check that no one is using fancy illicit footwork to titivate crumpled groundwork. He has to monitor each rig for 'extraneous attachments' which might yield an unfair advantage. He has to be dispassionate, honest and uninfluenced by personality or parochial favouritism.

While decisions may not be contested, the combatants demand to be treated with respect – especially the horse ploughmen. While participants in other sections of the game sometimes take a perverse delight in being scruffy, preferring to let their ploughs do the talking, the horseman (or horsewoman) looks to land 'best decorated' awards. Their plots may be smaller, their furrows may be shallower, and their speed of covering the ground may be positively glacial, but they are not above telling the organisers of an event how things should be done.

Tom makes it clear that he has nothing but admiration for those who try themselves against his exacting standards. The successful competitor is more than a mere robotic sportsman: he has to be able to work under time pressure, he has to have a good eye for a straight line, and he has to be something of a mechanic, with the working knowledge of a practical geologist.

And, it goes almost without saying, he has to be prepared to accept with good grace whatever the judges may decide.

Thanks to Tom Jenkinson of Gorey

18

THE EARLY DAYS OF COMPETITIVE PLOUGHING

The British brashly put forward a strong and earnest case for being to the forefront in the invention of ploughing as a sport and pastime.

Their assertions are received with little or no respect on the Dublin-Wicklow border, where the Rathdown Ploughing Society can make a convincing case for tracing its roots back as far as 1848. At a time when much of the population of the island of Ireland was starving, the folk in the barony of Rathdown were ready to rouse themselves from famine conditions to find the element of fun in the all-too-serious business of food production.

According to the lavishly produced history of the society, they were merely picking up the pieces of an even earlier foundation, the Wicklow Farming Society, which held matches over the period 1808 to 1843. Evidence for their claim is cited from a copy of *Saunder's Newsletter and Daily Advertiser* dating back to early 1849, where mention is made of the 'second annual Rathdown ploughing match and agricultural dinner'.

The event took place at Onegh, near Powerscourt, where the number of entries was considered by the *Saunder's* correspondent to be less than impressive, at twenty-six. On the other hand, the reporter was persuaded that those taking part exhibited a high level of skill in pursuit of the trophy on offer to the winner.

'While Rathdown may not, in fact, be the oldest ploughing society in Ireland, it is certainly well in the running for that title,' Barney Patterson, in his contribution to the history, modestly deduces.

Much of the interest in the matches of the time was generated by the lavishness of the prizes that were at stake. Those who were considering taking part in the society's promotion of 1850 at Windgates near Delgany must have been sorely tempted by the prospect of winning an iron plough, with five shillings going to the ploughman. Never mind the silver challenge cup.

Like horse racing and cricket, the sport was dominated by the big houses, where landlords presided over thousands of acres of the best land. Among the early patrons of the Rathdown Union Ploughing Society were Viscount Powerscourt (aka Mervyn Wingfield), Viscount Charles

Stanley Monck and Sir George Frederick John Hodson, third baronet of Hollybrook.

These men had the money to match their highfalutin titles. They had the horses, they had the best of equipment, and they had the most highly skilled ploughmen on their payrolls. And they liked to be seen to be associated with a pastime that was presented as being at the cutting edge of progressive farming.

They were to the ploughing scene what the Maktoum brothers or John Magnier is to modern bloodstock. The winner of the 1857 match, held in Bray, was Lord Monck, who took the iron plough, worth £3 and 10 shillings (or equivalent in other agricultural implements). His lordship was under no obligation to show his face at the field, of course. The dirty work was done by his ploughman, Donal Keenan.

Keenan, in fairness, was rewarded with £1 and 10 shillings – by no means a shabby return for a day spent in the field. Similarly, the runner-up spot was filled by Lord Talbot, who delegated the sweaty bits to his ploughman, James Buckley.

The titled gentry have since departed the scene. There is no patron of the society now, and the president (Robert Webster) does not even merit a 'Mr' in the list of committee members. Prizes in 2007, as a century and a half before, took the form of cash, with a twenty-litre can of oil for the best in each class.

History of the Rathdown Ploughing Society, researched and compiled by Tom Kehoe and Stanley Acres, 2008

19

PETER GRAHAM DIED! OH NO HE DIDN'T!

No words of the author are necessary by way of introduction, beyond stating that the following was extracted from an issue of *Saunder's Newsletter* published in 1849 by the authors of *History of the Rathdown Ploughing Society*:

> A report, ill advised as it is false, having gone abroad . . . relative to the consequences or the over-excitement of the men who competed for prizes at the digging match lately held under this union, I beg leave . . . to contradict authoritatively certain statements . . . It was stated that two or three of them fainted as they had finished their work; six of them

had to be blooded since the competition and others are reported dangerously ill.

[It is reported that] one of the diggers at the match, an able, well conducted, good labourer named Grimes [or Graham], has died from its effects.

Now the real facts are so much the reverse of this, that not only did not one of them faint, but they all awaited, and most anxiously, the decision of the judges, the call to answer their numbers and hear the awards read out; some, doubtless, diverting themselves in the tents during the interval; that only one instead of six of them was 'blooded since the competition'. Patrick Murray, he that won the second prize . . . and that not an individual of them is dangerously ill.

On the contrary, having seen most of them since in the service of their usual employers, they were most ready to express their thirst for another set-to, that they might show off their improved skill in competition with each other again, some observing that a perch was too little to try them on. Lastly, both the Grahams, one of who is said to have died from the effects of over-exertion, were at work next morning as early and well as usual, and are quite unaffected in their health, strength and spirits. I observed myself the lad, name Peter Graham, who was first done, because sick in his stomach, just as he had got the last sod turned; this, however, was occasioned by his having swallowed the pulp of an orange a few minutes previously.

The letter to the newspaper's editor was signed by 'John F. MacCartan, practical instructor with the Royal Agricultural Improvement Society'. He was enthusiastically convinced of the value of digging matches which were run in tandem with ploughing competitions. With ploughs and horses being beyond the resources of smallholders, it certainly made sense that they would be able to look after their land with hand tools. The practical instructor was convinced that such exercises were part of the drive to improve the yield from 'our exhausted fields'.

Despite the fact that the rumours of collapse and carnage were dignified in misleading public print, the practice continued. The year following the episode described by John MacCartan, in his efforts to set the record straight, his society once more advertised a similar match in Delgany for agricultural labourers 'who shall dig, in the shortest time and the best manner, one statute perch of land, not less than twelve inches deep'.

While the prize fund did not match the cash on offer to champion ploughers, the thirty shillings first prize was not to be sniffed at. The winner who emerged victorious from the stubble field set aside for the event was John Hand from Bray.

It took him just thirty-five minutes to turn over his plot, which measured eight and a half feet wide by thirty-three feet long. Patrick Murray of Ballybower laboured twelve minutes longer over his exertions and was slowest of the eight athletic fellows who took part. However, his style of spadework evidently appealed to the judges

(Andrew Kehoe, Esq., from Bray; Ulick Bourke from Crosses; and James Jones from Newcastle) and he was presented with the £1 second prize.

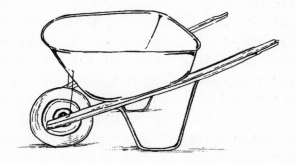

History of the Rathdown Ploughing Society, researched and compiled by Tom Kehoe and Stanley Acres, 2008

20

AN A-TO-Z

Automated calibrated feeders – solar powered, for sunny pastures

Bóthar – the organisation sends almost five hundred in-calf heifers to poor families each year.

Cattle crush – with maximum space for all your operating procedures. They should get one for James's Hospital, maybe.

Donohue Marquees – for all occasions. See also 'O' below.

AN A-TO-Z

European Commission – with special reference to climate change and biodiversity.

Fexco – low-cost private client stockbroker.

Grease traps – and Molloy Precast also do storm-water attenuation.

Horse Passport Agency – donkeys may also apply.

Irish Guide Dogs – not only for the blind but also for families of children with autism.

Jetta – with 527 litres of boot space.

Knock Hill Billys – on the bandstand.

Land Rover Terrapod – showing how to drive very, very slowly down a very steep hill.

Mountbolus pipe band – not so much buffaloes as puffaloes.

Neven Maguire – seared sea scallop with black-pudding sausage, anyone?

O'Donovan Marquees – for all seasons. See also 'D' above.

Pistachios in the Park – they don't grow pistachios round Portiuncula – or Prosperous, for that matter.

Quick Tag Limited – all the way to a ewe near you from Ballycastle in County Antrim.

Revenue's Custom Service – meet Mickey and Lulu, the drug sniffer dogs.

Ssang Yong – 4x4 Korean style.

Tote – betting on course, on the phone or online.

Urban Forest – equine bedding.

Vendeen – that's a breed of sheep, big in Nobber.

Welly4U – thank goodness someone in Navan has a sense of humour – and a big pair of feet.

Xtract Energy – extracting oil from shale.

Ying's Bags – just the job from Dublin 4, hand-made in silk.

Zero Grazing Systems – keeping the cows in the cow-house.

AN A-TO-Z

Question: What do all the above have in common?

Answer: They all featured in some shape or form at the National Ploughing Championships at Annaharvey farm near Tullamore in 2007.

Wonder: what on earth do Mickey, Lulu and the Revenue Commissioners have to do with keeping a straight line in a stubble field?

21

ROMANTIC IRELAND DEAD AND GONE? NOT A BIT OF IT!

On 5 August 2007, more than four and a half thousand 'vintage' tractors were assembled in a seventy-acre field at the base of the Cooley Mountains. They were variously brought to the venue on lorries, on trailers or under their own steam from just about every corner of the island (including Gweedore), with a few welcome strays from even further afield.

Every last one of the vehicles which arrived in County Louth in response to John Hanlon's invitation was of pre-1977 origins. Some were immaculately presented, refurbished, restored and re-invigorated, while others looked as though they had just been rolled out of lengthy and unloved hibernation. Polished or down at heel, they were all verifiably thirty-plus years old.

It was no surprise that there were Fords and Fergusons aplenty, for Ford and Ferguson were the dominant marques supporting food production in Ireland for more than the span of a generation up to 1977. By no means did those names have a monopoly, however, with makes such as Perkins, John Deere, McCormack, Nuffield, Steyr, Bates, David Brown, Leyland and International sprinkled among this most arresting of convoys.

The event proved, if nothing else, that tractor manufacturers used to have very little notion of colour beyond a stock palette of grey, blue and red. The odd white charger stood out in startling contrast, even in the murky light of a wickedly wet weekend.

Legend has it that three inches of rain fell during the twenty-four hours up to the starting time of 3.30 PM on that memorable Sunday afternoon. And who are we to query legend? Certainly, portions of the allotted field were reduced to the consistency of milky-brown liquid chocolate by the downpour.

The first machine to be registered for the event was brought from England by Cooley regular John Moffitt. It was also the oldest of the lot: a 1903 rarity, an Ivel, commonly accepted as the first practical small tractor ever put on the general market. The honour of leading off the procession was handed to Dunleer man Gerry King, whose hulking 1919 Avery – a massive pre-World War Two American import – loomed large over all the Ferguson minnows.

The plan was to re-establish Cooley as the holder of the world record for vintage tractors all working in the same place at the same time. Hanlon and his committee first held the honour in 2002, when they attracted 1,832 old tractors to their field to earn their place in *The Guinness Book of Records*. They were ready to welcome around eight hundred on that sunny occasion, but more than double the expected number turned up. They smashed the existing mark of four hundred or so, set in South Africa.

It was a matter of good-humoured affront that the baton was snatched from Louth hands by an Australian effort that rallied 1,901 tractors, all working simultaneously in the one field in 2004. Then, a year later, it was the turn of the Wiltshire folk at Hullavington near Malmesbury in England: they raised the bar to 2,141.

The ball was in the Cooley court. Stern action was required, and the challenge captured the imagination of vintage enthusiasts from Banbridge to Bandon. Not even the three saturating inches of precipitation and knee-deep brown slop was going to prevent the Irish regaining their grip on the record.

As it turned out, the 2007 event attracted at least two tractors for every one that performed at Hullavington. Approximately fifty frustrated owners were turned away from the registration marquee, told that they were too late. And a score of engines were too wet or too old to start on the command – but that still left 4,572 tractors in action.

They looked like something from a battle scene in *Lord of the Rings* – the difference being that, in the movies, the producers use computer trickery to give the impression of vast numbers. In this case, the numbers really were vast, without any codology. As the showers cleared, amidst much smoke and noise, the 4,572 ploughed and harrowed their way across the sodden turf to re-establish Cooley's pride.

www.farmersjournal.ie/2007/0811/farmmanagement/machinery/index.shtml

Cooley Vintage Festival, World Record 2007 official DVD, written and narrated by George Dallas, Linton Film Productions

22

'No Way Would I Go Back to Ploughing'

'No way would I go back to ploughing.' That's Isaac Wheelock. Isaac has hundreds of acres under cereals. Wheat is his favourite. Yet he made his plough redundant shortly after the turn of the millennium. So Isaac must be a raving heretic. Or is he?

The way he tells it, Isaac is not so much tearing up the rulebook of good farming practice as going back to an era when soil was treated with greater respect. He recalls the days of his great-grandfather, when a merry man with a pair of horses would set out with a smile and a song on a sunny day to plough an acre. It had to be a sunny day (or at least dry) because otherwise the horses would slip in the mud and the job would become impossible. The merry man with the horses could go down no deeper than four, maybe five, inches below the surface – else the poor gee-gees would go on strike or give themselves hernias.

Move on a few decades to the arrival of the nippy little grey Fergie. 'The best tractor ever made', as Isaac fondly calls it, had no four-wheel drive, no differential lock to keep the wheels turning regardless of conditions underfoot. The result was that the ploughman on the grey Fergie also had to sit on his hands rather than work in the wet. But times, and technology, changed. Farmers became more specialised, and the weapons in the arsenal of the grain-grower progressed in calibre from popgun to Big Bertha inside the space of a generation.

By the time Isaac Wheelock joined the peace movement, humungous brutes of machines were in the field. These muscular heavyweights are capable of hauling ploughs that turn five sods at a go. Each sod may be twenty inches wide, and each furrow is deep in proportion. The work may be done in practically any weather, because Big Bertha does not suffer hernias. She has more horsepower under her bonnet than features on the entire card at Epsom or the Curragh come Derby day.

Big machines mean compaction of the soil, Isaac points out. Yes, the mega-ploughs are ruthlessly efficient at burying weeds, and the job is done in double-quick time. But the worms that used to assist in aerating and fertilising the ground have been largely squeezed away. And there is an alternative approach – just ask Isaac. It's called min-till (minimum tillage) or non-invasive tillage or eco-tillage. No ploughing.

The idea is that weeds are controlled by herbicides and that the ground is prepared for the planting of seed by a good scuffling, down to a depth of no more than four inches. This has the advantage of requiring a great deal less diesel and far fewer man-hours than traditional ploughing – no small consideration, when you have at least eight hundred acres to tend. His neighbours may laugh when they peer over the ditch to spy Isaac's seedlings come up sad and sickly but, after the slow start, his crops are well up to standard by harvest time.

He says he is using maybe a fifth less fertiliser than he used to scatter when he ploughed. His son Andrew says that his soil is definitely 'fluffier' (a charmingly descriptive turn of phrase), as it begins to breathe after years of compaction. The da notes that the worms are returning.

Instead of a plough, he must spray Roundup, at a rate of one litre per hectare, twice in the twelve months. In other European countries, such as Spain and Germany, governments happily subsidise landholders who opt for min-till, or eco-till, or non-invasive tillage. In Ireland, as he notes with a resigned shake of the head: 'People don't change.' Min-till is the system of cultivation preferred on no more than 35,000 acres of grain.

Fewer than one hundred growers in Ireland use non-invasive tillage. They believe there will be many more in future – and much less ploughing.

With thanks to Isaac Wheelock of Davidstown

23

THE WEATHER MAN

The English have a phrase: 'Mustn't grumble.' The Irish have no such expression – certainly not Irish farmers. For them, grumbling is an art form – an art form that starts at the top.

For more than forty years, that distinguished Galway native the late Michael T. Connolly compiled his annual report on the state of farming in County Wexford. In all of the four decades up to the time of his retirement as chief agricultural officer in 1977, he never came across a year of good weather.

As a young agricultural instructor, he recorded how weather conditions in 1936 were the most unfavourable in years. That year, a wet winter meant that seasonal farming operations such as ploughing were severely hampered –

and nigh-on impossible in low-lying areas or in holdings with heavy soil.

Things did not pick up in 1937, as it was only with difficulty that land was prepared for the sowing of crops in time. It was no better in 1938, when weather conditions were once more described as unfavourable. Nor was 1939 much of an improvement: that year, the ploughing and sowing of winter cereals was hampered to such an extent that it was impossible in low-lying fields.

Strangely, for once Michael T. made no mention at all of the weather in 1940. However, as he reviewed 1941, he noted that the sowing season was late and 'unfavourable' (his favourite adjective, along with 'unsuitable', when dealing with matters meteorological). Despite this, crops gave very satisfactory returns. There was more of the same in 1942, though once again, the wet growing conditions did not prevent very satisfactory wheat yields.

The year 1943 was unfavourable (again) for cereal crop production. The reader almost begins to wonder why anyone bothered venturing on to the land. It was the same downbeat story in 1944, when the weather, as reported by Michael T., was in contrary mood and unsuitable for crop production, as six dry months were succeeded by six wet months.

The rain-soaked spell was broken in the early weeks of 1945, when, from 2 January to 29 January, there was heavy frost practically every night. Then normal service was resumed: February was entirely wet, causing floods in many areas. And so it continued, year after soggy, unfavourable, unsuitable year. An inspirational man who was in other respects so positive, progressive and open to innovation, Michael T. seems to have taken each misplaced shower not only as a surprise but also as reason to don the sackcloth and ashes.

The spring of 1959 was by no means dry or suitable for soil cultivation. The year 1960 was an unusually wet one. The quality of wheat grain appeared quite good, but much of it was deemed unmillable because of incipient sprouting. In mid-September 1961, conditions were 'ideal' (steady, Michael) for harvesting crops, but then the hurricane of 16 September lashed corn. Ripe barley and wheat were broken down, deadheaded and shedded, with more storms on the way.

The trouble with 1962 was that the rainfall came at the wrong season – i.e. in July, August and September. During January and February of 1963, the weather was the most severe that had been seen for those months since 1947. The gloom continued throughout the rest of Connolly's career. In 1975, he adverted to the severe drought, which extended from March to mid-July. Grass and crops, especially potatoes, were severely retarded. The drought of 1976 severely affected the growth of cereals and root crops. The wonder is that farmers bothered growing them at all.

Michael T. Connolly retired in 1977, so it was left to his successor to chart the triumphs and vicissitudes of that year. With the words 'Against the odds of a wet and very late spring . . . ', a noble tradition of glum weather-watching continued.

Forty Years of Wexford Agriculture by Michael T. Connolly, edited by Richard Doyle

24

A MINORITY PURSUIT

Ireland is not the bread basket of western Europe by any means. Grain growing may be portrayed as an incidental part of the agricultural scene in a country that devotes much more space to its cattle, whether for dairy or beef. Yet there are sections of the island that miraculously turn into a patchwork of brown in amongst the green as the plough does its work, where the countryside takes on a deep gold colour at harvest time.

Other parts are not so lucky. The area devoted to corn in thirteen counties – Galway, Tipperary North, Westmeath, Kerry, Mayo, Roscommon, Limerick, Monaghan, Cavan, Longford, Clare, Sligo and Leitrim – combined (no pun intended) is scarcely more than in County Wexford on its own. You may then add in the heroic efforts of grain growers in Offaly and most of Donegal to match the acreage under corn of County Cork – the biggest producer of the lot in the Republic. The figures from Central Statistics show Cork in first place, followed by Wexford, Meath, Kildare and Kilkenny, in that order. For every acre of corn grown in Leitrim, in the reference year of 1980, Cork had 816 acres.

The balance (or imbalance) is reflected at the fascinating Museum of Country Life at Castlebar in Mayo. More space is devoted there to the traditional costumes of Aran Islanders and the way they dressed their boys as girls than to the business of ploughing. Indeed, there is just one plough on display there – a lovely ancient timber creation that could pass as a sculpture, remnant of an age when farmers did their inventive best with whatever resources were to hand.

For a celebration of all things tillage, the place to go is at the other end of the country. The symbol of the Irish Agricultural Museum at Johnstown Castle near Wexford is a plough. The collection was initiated by the passion of Austin O'Sullivan and now continues to delight and enlighten under the curatorship of Sharon Quinn. Here, Ulsterman Harry Ferguson is given the pride of place that is rightfully his.

One wall of the museum is dominated by a huge painting that shows the two H.F.s, Harry Ferguson and Henry Ford the First, giants of the industrial epoch. The pair are pictured sitting at a table outdoors at Ford's farm in Dearborn, Michigan, exploring the details of a tractor model. The message is clear: that the man from Dromore was hugely successful as a businessman and enormously influential as an agricultural pioneer. Credit is extended to Ferguson's design team of Willie Sands, Archie Greer and John Chambers.

The Johnstown collection features several of the Ulsterman's tractors, including a rare Ferguson Brown Type A, the 547th of just 1,354 that came off the production line in Huddersfield in the four years up to 1939. It resembles a length of sewerage piping with added wheels, seat, engine and grille – and it continues to leak oil to this day. Type A enjoyed limited commercial success, as the asking price of €284 was about double that of a Fordson. It was only after the war that the little grey Fergie – so coloured to merge with the landscape – became an essential part of that same landscape.

Examples from the era before engines were commonly available are also on show at the Agricultural Museum, where there are seven ploughs which were made for use with horses. One of the seven has been dismembered in order to show and name the individual parts. The set includes an oddity, made at Pierce's foundry: a four-sod seed plough which gave the earth a shake as it dropped in the seeds. The notion was briefly popular in the early 1900s, and it may be no coincidence that the example here shows little sign of wear.

25

ACRES AND ACRES

Wikipedia, the cheating researcher's internet standby, lists no fewer than seventy-seven units of area – more if you throw in all the Romanian ones, of which there are plenty.

At the invisible end of the range is the 'shed', deployed in nuclear physics to measure the cross-sections of subatomic particles and described as extraordinarily small: it takes all of 10,000 sheds to make one square yoctometer. And you could step on a yoctometer without fear of tripping. Contrast such minuscule-ness with the 'section', which covers a square mile of wide-open territory in the United States, according to Wikipedia.

There are plenty of handy units of land, with the humdrum acre leading the pack. Unfortunately, many of those that pop up on the cheat's computer screen are given no credence by the musty, dusty authority of respectable, printed authority. *The New Oxford Dictionary of English* has no room for the mysterious 'bunder' (alleged to be one-twelfth of a 'mansus' in parts of the Low Countries), or for the 'jerib' (said to weigh in at around half an acre in Afghanistan). Nor can the editor find it in her heart to authenticate the 'feddan', plausibly put forward as meaning 'a yoke of oxen' in Egypt and environs. Pleas that the oxen in question can churn over 4,200 square metres in a day's labour fall on deaf ears amidst the dreaming spires of the university city.

Similarly, there is no comfort for devotees of the 'pari' in the pages of *Chambers Dictionary*: cheat's claims that it is roughly equivalent to a hectare down Minipur way in India simply do not register in that quarter. Though it may be sniffy about many of the numerous measures of real estate on the subcontinent, *Chambers* shows its Edinburgh roots by standing over a goodly share of Scottish contenders: the davoch, the groatland and the oxgang, for instance.

The same dictionary also lays out in cold print that the obsolete Irish acre (7,840 square yards) and similarly redundant Scottish acre (6,150 square yards) were appreciably bigger than the 4,840 square yards statute acre, which has been in vogue for at least a century and a half. Both *Chambers* and *Oxford* concur that the word 'acre' comes from the Dutch or German word for a field (*akker* or *acker*, respectively). The word's roots appear to stretch back even further to the Sanskrit *ajra*, meaning a plain. It is universally accepted that the acre was originally the amount of land a yoke of oxen could plough in a day. How the ancient Irish and Scottish oxen managed to be so much more productive than the English beasts is not explained.

The metric rival to the acre is, of course, the hectare, which sprang not so much from the soil as from the sea-green incorruptibility of the French Revolution. The logic of the revolutionaries proclaimed that the metre was a fraction (one ten-millionth, to be precise) of the distance from the Equator to the North Pole. From there, the obvious next step is to come up with a unit of area that is 10,000 square metres. Add the veneer of some spurious Greek derivation – 'hekaton' is 'one hundred' to classical scholars – and you have the world's standard unit of land measurement. *Oxford* looks down its nose at this artificial concoction, reckoning that the adaptation of 'hekaton' is irregular in such a circumstance. There are several earthier, dictionary-approved alternatives to hectare or acre, such as the Hindi *bigha*. It is (ahem) big, ah, in northern India, covering anything from half to two-thirds of an acre, if you believe *Oxford*, or one-third to one hectare, if it is in Chambers that you trust.

Both dictionaries like the look of the 'morgen' – derived from the Dutch or German for 'morning'. *Chambers* suggests, therefore, that a morgen must represent a morning's work. *Oxford* is of the same mind but offers a very ambitious reckoning of what could be achieved in the half-day shift: two acres in Norway, Denmark and Germany.

But the dictionaries draw a blank on Wikipedia's 'stremma' (supposedly 1,000 square metres of Greece), 'lugar' (allegedly 7,166 square metres of Transylvania) and 'manzana' (possibly 1.68 acres of Central America). Manzana? A likely story.

www.wikipedia.org

New Oxford Dictionary, edited by Judy Pearsall, 1998

The Chambers Dictionary, tenth edition, editor-in-chief Ian Brookes, 2006

26

'WE CAN BE HEROES'

> 'We can be heroes, just for one day'
> DAVID BOWIE

Just for one day? Wrong! The heroic tillage farmers of Ireland saved the nation and carried the Republic through the hard times of World War Two. The Emergency lasted considerably longer than just one day.

In his report on the state of Wexford farming for the year 1935, agricultural officer Michael T. Connolly (no mean hero himself) noted the success of the State promotion of wheat growing. By the simple ruse of paying producers a few extra shillings, they had been induced to increase the amount of land devoted to the crop from a miserable 21,000 acres across the Twenty-six Counties in

1932 to a magnificent 163,473 acres just three seasons later.

'It will be seen, therefore, that taking the country as a whole, we produced nearly eight times more wheat in 1935 as in 1932,' calculated Michael T. with a happy chuckle. 'The past three harvests have proved that we can grow wheat as successfully as any country in the world.' At that time, the impulse behind paying 23 shillings (and sometimes more) for a barrel of wheat was to replace foreign grain with home-produced foodstuffs for the benefit of the national trade balance. As Marie Antoinette might have observed, if citizens were to eat cake, let them eat Irish cake.

When war broke out in 1939, the notion of import substitution became not so much desirable as absolutely imperative, since there was no one selling grain to Ireland. By then, the area under wheat had risen to a very creditable 225,280 acres, but Connolly calculated that it would require 700,000 acres to keep the nation happy. This was a tall order. Running his rule over all the tillage crops, the Galwegian reckoned that 60 percent more land should be under the plough.

The compulsory tillage scheme was introduced; this required farmers to make at least an eighth of their land arable. This was no big deal in the ploughing heartlands such as Cork and Meath, but it was a lot to ask elsewhere. Many years later, cattle and sheep farmers were still muttering about the deleterious effect of the disruption to their precious grassland in the seventies. Still, it had to be done, and the fact that even animal feed grain prices leaped 50 percent in the first three months of the Emergency sugared the pill.

The area across the Republic that was ploughed rose by very nearly a quarter in 1940, but Michael T. reflected the mood of higher authority when he wrote: 'There is no excuse for the half-hearted response given during the last season by certain owners of large grass farms very suitable for tillage purposes. . . . Our chances of obtaining wheat from outside sources must be looked on as very remote.' In 1941, he reported that the wheat acreage sown was the largest since 1846, while the acreage under crops for animal feed was the largest ever recorded. By now Wexford was devoting 34,056 acres to wheat alone – considerably more than the entire State had managed just nine years previously. Growers in the Model County (and Kilkenny) were rewarded with high yields. Still, the call for food production to be increased further was handed down from above: 'Those who are in a position to judge, anticipate and are preparing for a long war.'

While there were national surpluses of meat, milk, butter and eggs, the lure of 45 shillings a barrel was still not enough to persuade landholders to produce all the wheat required. A scarcity of fertiliser did not help. Nevertheless, the area under wheat more than doubled in 1942, bursting through the half-million-acre mark. As Connolly noted proudly: 'Farmers in this country have completed their third production season under emergency conditions. During that time much has been asked of them: much has been given. Few except those closely related with the land realise the enormous changes, almost revolutionary, which have taken place in agricultural production during the short period.'

This salute to farmers was carried in the 1942 report before an exhortation to yet greater efforts. 'FURTHER INCREASE NEEDED' was the headline; 'An increase of 75,000 acres of wheat is still required' was the demand. No problem. We can be heroes: by 1944, the figure had risen yet again, to 642,000 acres.

> Yields from sowing corn have increased dramatically over the years. The Irish field that gave no more than three-quarters of a tonne of barley or wheat per acre in 1898 produced over three tons of grain in 1977, according to Michael T. Connolly, chief agricultural officer for County Wexford. Or, to look at things another way, the average hectare of wheat around the world gave a yield of 1,057 kilos in 1955. By 1985, this had more than doubled, to 2,204 kilos. Some places made more progress than others. The rise in Africa over the period was from 720 kilos to 1,271 kilos. Meanwhile, the top of the class was the Netherlands, which reported 6,649 kilos per hectare in 1985, compared with 3,923 kilos three decades previously. The figures for Ireland were well up with the Dutch, rising from 2,799 to a commendable 6,600 kilos.

27

THE ORIGINAL CELTIC FIELDS

Barley and wheat make up at least three-quarters of modern Irish grain cultivation, with room for some oats and maize. The farmers of early Christian Ireland appear to have had a more varied roster of crops, with rye and spelt also on the list. Wheat was highly prized, not least because it could be used to make beer as well as porridge and bread. However, it was also the most problematical of plants. Developed from a weed that sprang up in the heat of the original grain belt in ancient Mesopotamia, wheat was reluctant to thrive in the dampness of rain-washed Ireland.

Rye really was a better bet. It could be sown in autumn. It was hardy and of long-established economic importance. Evidence from a site at Carrowmore in Sligo suggests that it was cultivated on the island as long ago as the Bronze Age. Though rye was widespread, the archaeologists have never found anything to suggest that it was ever truly popular. Anyone who has ever been force-fed black, leaden pumpernickel can readily understand why. Saint Finnian, founder of the great monastery at Clonard in County Meath in the year 520 AD, appears to have passed on the Ryvita when selecting his diet. Clearly a man of refined taste, his worship dined on barley bread during the week and then allowed himself a portion of wheat bread come Sunday.

The business of raising animals, not least goats, was a much bigger enterprise than tillage for the folk of seventh- and eighth-century Ireland. Yet cereal production was economically important, and a sophisticated activity. Those who practised the arts of corn production were fully aware of the benefits of fertiliser, for example. Cattle poop was considered the best manure, so there was an established trade in cow and ox dung.

Growers not only had to contend with weeds, in the absence of Roundup, they also had to keep flocks of birds and herds of wild deer at bay. And of course they were at the mercy of the weather. In 773, yields were devastated by drought. There are unsubstantiated suggestions that a plague of locusts descended on Ireland some time late in the ninth century. The corn crop of 1012 was wrecked by heavy rain, while the Annals of Innisfallen record that fields were flattened by gales in 1077. Such events triggered real hardship among a population which had no recourse to Argentinian imports.

The cutting edge in pre-Christian, ox-powered ploughing in Ireland was provided by timber, bone and stone. Indeed, there is little or no evidence of iron plough shares being used in the country before the seventh century AD. Such primitive equipment hardly ruffled the surface of the ground, barely turning a sod. Accordingly, a second run was required, at right angles to the first pass. The result was small, squarish areas of cultivation: the original Celtic fields.

Four oxen was the preferred strength of a team, and it was not until the thirteenth century that breeds of horse that were up to the task began to take on much of the work. A well-to-do farmer was expected to have his own set of ploughing gear, and he had to run to six oxen in order to be able to carry out his corn-growing duties. The lower orders had to make ox-pooling arrangements – an early example of the cooperative spirit in Irish agriculture. Indeed, much of the preparation and cultivation of the land must have been undertaken in small plots exclusively by manual labour.

Occasionally, a higher power lent a hand. Scholar Fergus Kelly retells the story of a sceptical Columb Cille arriving to visit Bishop Etchen, to find his host ploughing. The bishop's rate of progress around the field continued uninterrupted when Columb Cille removed his ploughshare. When the guest took away one of the team of oxen, a wild stag miraculously appeared from the nearby woods to fill the vacancy.

> Fat hen, which is now considered a menace when it pops up among stands of grain, was a prized crop during the Stone Age. It is a precursor of modern-day spinach and cabbage.

Early Irish Farming by Fergus Kelly, School of Celtic Studies, 1998
www.monasticireland.com/historicsites/clonard.htm
The staff in the library at UCD
James Brennan – at least he tried

28

THE DIET OF WORMS

> 'And though the holes were rather small'
> THE BEATLES

Legendary naturalist Charles Darwin is famous for his pronouncements on evolution. However, the great man was not concerned only with the headline-grabbing tale of how ape became man. Shortly before his death in 1882, Darwin published a work called *The Formation of Vegetable Mould Through the Action of Worms*. The book sold in its thousands, and demonstrated the author's gift for alighting on topics that were both important and fascinating.

In the book, Darwin stated his belief that the black topsoil in the fields of his native Shropshire comprised worm castings. Yes, worm poo. This superb growing medium was made up of faecal matter passed through the primitive guts of thousands upon thousands of worms. Darwin's critics wondered how such a weak vegetarian creature could produce enough material to raise the level of a field by several inches in a few short years. Yet his conclusions, backed up as usual by meticulous observation, were correct and are generally accepted to this day.

The nature-loving public may become agitated about the perilous predicament of the corncrake, or the poor prospects for the panda. But the panda and the corncrake are but nature's sideshows compared to the humble earthworm, whose activities make him (or her: earthworms are hermaphrodites, after all) a significant ally of the farmer and, through the farmer, of all mankind. We may weep briefly if we lose the corncrake; we may starve if we lose the earthworm.

It is estimated that there are about three thousand species of earthworm. They may be found just about anywhere that is not desert or permanently iced over. They are slow, too, to populate areas where glaciers recently flowed. Their poo is found practically everywhere, and worm poo is an essential ingredient of arable farming. It is the worms, after all, which replace the eroded material that is washed away in the wake of ploughing.

However, their contribution goes much further than merely bulking up the soil. They also tend the very top layer of the earth's crust, aerating it, fertilising it and draining it – as assiduously as any gardener. Each worm may be small but, with one hundred or more individuals in a typical square metre of tilled ground, their combined impact is colossal. They are impressively resilient too, sometimes living up to ten years – if they are not gobbled up by gulls following the plough.

Scientists have demonstrated again and again that the presence of worms is good for agricultural productivity. The worms break down the remnants of past crops into material that feeds future crops. According to one study, worm casts are thrown up at a rate of at least three kilogrammes per square metre of farmland per annum. Upland terrain with few or no worms often has a layer of undigested, un-decomposed vegetation, which is no good to man or beast. The presence of worms is an indicator of soil health.

Worms boost the levels of nitrogen – the essential building block for all plants – without recourse to artificial manures. It is thought that they may also affect the availability of aluminium, calcium, iron, manganese and a basket of trace elements. The scientists urge farmers to work more closely with the miniature, mobile, living biochemistry set that is the wonderful earthworm.

Their burrowing lifestyle, leaving a maze of little tunnels, has the effect of improving field drainage. This means that water tends to soak into the soil, like a sponge, rather than running off into the nearest gripe in a brown stream of dirt and nutrients. The helpful creatures also pull artificially applied fertiliser and pesticide down below ground level, so that they are less likely to be washed away.

The warren of worm burrows provides a ready channel for easy root growth of cereals.

Dutch experiments have suggested that the presence of a dense population of earthworms can improve the quality of wheat, as well as improving grass growth in pasture. Farmers in various countries around the world have imported worms to help them in the work of food production. North America gave Europe the pest which is the grey squirrel. Europe, in return, has posted various members of the hard-working *lumbricidae* family to shore up North American agriculture.

No argument as to who has the better of the deal.

Earthworm Ecology and Bio-geography in North America, edited by Paul F. Hendrix

http://charles-darwin.classic-literature.co.uk/formation-of-vegetable-mould/

29

FULL STEAM BEHIND

'Polly put the kettle on'
TRADITIONAL

It was not until the 1950s that tractors became commonplace in the fields of Ireland. And the original drive towards mechanised tillage had come not so much from Europe as from the United States – a hundred and more years earlier. Ambitious landholders were opening up large expanses of broad prairie throughout the nineteenth century, and they needed power to make the most of their holdings. They wanted to grow corn on the largest of large scales rather than raise cattle, and they needed very little persuasion to try motorised methods.

It was steam that led the charge towards making the prairies truly arable. As early as the 1830s, the farmers were stoking up the boilers on steam-powered reapers. By 1860, the firm of Pitt Brothers alone had fifty workshops dotted around North America making threshers. These threshers were normally powered by portable steam engines, which were commonplace across the country.

The business of ploughing proved to be a tougher technical challenge. Still, as early as 1855, one Obed Hussey of Baltimore announced that he had developed a steam plough. Three years later, Hussey's design was overtaken by an effort from John Fawkes of Christiana in Pennsylvania. Bison beware: their grazing grounds would soon be at the mercy of the grain producers.

An English firm called Fowlers were also on the case. Amidst much hullabaloo, they presented a plough which ran on a cable between two steam engines. This looked well in the illustrated prints, but it was a costly and over-elaborate solution to the challenge. Such devices were adopted by some sugar-beet growers on the West Coast of the United States, but they never took off on the broader plains across the middle of North America. They were seen as being too finicky.

A more practical approach was required, and it was not long in being delivered. In 1868, the first Standish rotary steam plough, christened the Mayflower, was built in California. It may have been standish by name, but not by nature. The beauty of it was that the Mayflower could move, covering an impressive five acres in a day's ploughing, with a reputed 60 horsepower at its command. In the same year, Redmond's steam plough was launched in New York. That must have been some launch: this not-so-little baby weighed in at 5,000 pounds.

There was a ready market for such big beasts. The smoke from their solid-fuel engines was soon drifting across Dakota, Colorado, Montana, Nebraska and the rest of the US grain belt, not to mention western Canada. By 1900, there were thirty factories churning out large steam tractors, and claims were made in Saskatchewan that one such had turned over 160 acres in a period of just twenty-four hours. Spectators at the Brandon Manitoba fair of 1909 were treated to the spectacle of one acre being ploughed in precisely 7 minutes 58 seconds. There's never a McWhirter around when you need one.

Given the lumpen tonnage of these mighty pieces, the makers went to all sorts of lengths to spread the load, as the machines were operating on soft land. In the era of steam tractors, ingenuity of design was prized above elegance. Some of them looked like the innards of vast cuckoo clocks, they boasted so many cogs and wheels. Others resembled artillery pieces primed to blast the land into hapless submission.

They were dinosaurs: expensive, heavy and tricky to operate. The vested interests behind steam resisted gamely, but a more efficient, nimbler species was on the way. The head of the Kinnaird-Haines corporation stated that he would not allow his steam beauties to share a freight car with the competition in 1910.

It was a defiant last kick of a doomed species. The gasoline tractor had arrived.

The Agricultural Tractor 1855–1950, compiled by R. B. Gray, American Society of Agricultural Engineers, 1975

30

THE FIRST TRACTORS

So, who had the honour of presenting to the world the first plough-pulling tractor with an internal combustion engine? Good question. The credit should not go to the Charter Gas Engine Corporation. They constructed six gasoline 'tractors' in 1889 for use on farms in the northwest of the United States but the engines used were of the stationary type, and were not capable of working their way around a field.

Perhaps the world should know more about George Taylor from Vancouver, on Canada's Pacific coast. He patented a 'walking type motor plow', powered by petrol, in 1890. If it was built at all, it was probably made on a very small scale, fit for the garden rather than the rigours of agriculture.

THE FIRST TRACTORS

From the early 1890s, William Deering and Co. experimented with corn pickers and mowers – but no sign of a plough. In 1892, John Froelich (yes, he of Froelich, Iowa, fame) completed a gasoline-driven tractor that was capable of threshing in a range of temperatures. Froelich's ingenuity inspired the establishment of the Waterloo Gasoline Traction Engine Company (later acquired by John Deere) but the man himself moved on to inventing other things, like dishwashers and air conditioners, without ever applying his great mind to how it might be possible to pull a plough behind one of his machines.

C. H. Dissinger & Brothers of Pennsylvania, J. A. Hocket of Kansas, the Lambert Gas and Gasoline Company in Indiana, the Davis Traction Company of Iowa, and Huber Manufacturing Company in Ohio – to name but some of those involved in the business – all dabbled in this area in the 1890s. But it was not until 1902 that progress from 'belt' work such as threshing to fitting a draw bar for ploughing and harrowing was made.

Step forward C. W. Hart and C. H. Parr of Charles City in Iowa. The place was apparently given its name in their honour, for they were a proper pair of Charleses. They put the city on the map when they arrived from Wisconsin and made it the headquarters of the Hart-Parr Gasoline Traction Engine Company. According to local lore, not only were they manufacturing one-third of the world's production of tractors in 1907, they were also responsible for coining the very word 'tractor' in its present sense. Credit for this was given to their sales manager W. H. Williams. (With initials like that, we know that he was not a Charlie, proper or otherwise.)

Hart, Parr and Williams did not lack for rivals. Just as they delivered their first model, in the same year, 1902, the Massey-Harris Company in Wisconsin gave birth to the 'Wallis Bear', with a four-cylinder engine, capable of tugging ten fourteen-inch ploughs around the place.

The pace of innovation and development was feverish. Six years later, at the Winnipeg Industrial Exhibition of 1908, no fewer than seven vehicles were submitted for scrutiny in the 'light agricultural motor' competition. Neither the Charleses nor Massey-Harris were represented. The International Harvester Company entered three machines, ranging in horsepower from a handy fifteen to a hefty forty. Also in contention were machines from the Transit Thresher Company, Kinnard-Haines Co. and Marshall Sons & Co.

The seventh entry came from Universal, but it was 'withdrawn owing to breakage'. The winner by a whisker, all the way from Minneapolis, turned out to be the Kinnard-Haines, which used 75.5 pints of gasoline and 48 pints of water in performing the set routines. It ploughed an acre in 37.3 minutes.

The seven in 1908 became thirteen in 1909, as the competition heated up. Avery, Russell, JI Case and the Gas Traction Company entered the fray. The machines ranged upward in price from the $1,700 for International

THE FIRST TRACTORS

Harvester's baby tractor. Steam's days were now well and truly numbered, although as late as 1920, fifteen concerns manufactured a combined total of 1,766 steam units in North America.

www.froelichtractor.com/tractor.htm
www.masoncitynet.com/charlescity/story_template.php?audio=03.txt

31

DOWN BY THE BRAHMAPUTRA

Bangladesh has it all. Three-quarters of its land is fit for ploughing. The soil delivered by the Brahmaputra from the Himalayas is among the richest in the world. The climate is warm and there is no shortage of water. Spit out a tomato pip practically anywhere in Bangladesh and, within a few weeks, a fresh crop will be ready for picking. It should be the Garden of Eden for the new millennium. Everyone needs to eat, after all.

In fact, Bangladesh is anything but a latter-day paradise. When it measured the wealth of nations in 2000, the *Economist* found that, among the sixty-seven major nations for which it provided data, only Nigeria scored worse in terms of gross domestic product per head of population. The magazine's statistics suggested that there is in fact very little (if any) link between fertility in the fields and economic success on the global stage.

Indeed, Bangladesh's most valuable export in the year in question was not tomatoes, or rice, or any of the other crops that sprout so readily on the banks of the Brahmaputra. Close to three-quarters of the country's export earnings came rather from the sale of clothing, run up by the nimble fingers of the country's textile workers. The Bangladeshis may not be the richest people in the world but they are by no means the most dowdily dressed. There is a thriving market in Dhaka for designer-wear surplus and seconds.

In a crazy world, where some people go hungry and others seriously consider good wholesome grain as a potential fuel for their central-heating boilers, the potential for growing food is a poor indicator of material well-being. For instance, 60 percent of Denmark is rated as being of arable standard, compared with 12 percent of Japan. Yet in each case, no more than one in every twenty Danish or Japanese workers is engaged in farming. On the GDP-per-head measure, the two countries were very closely ranked in 2000, with Denmark, on $33,000, beating Japan, on $32,000, by a nose. Denmark makes its money primarily from 'manufactured goods', while the Japanese stream of earnings is led by the best-built cars in the world – and not just Toyota.

Less than a fifth of the territory of the United States is fit for the plough, yet this did not prevent the Americans earning $29,000 per capita in the year in question.

Wealthiest of the lot were the Swiss, on just under $40,000. They labour under the handicap of having just 12 percent of their land as arable – similar to Japan, Iraq, Mexico and Ireland. Agriculture is barely a flicker (2 percent of GDP) among the vital signs of their economy, Gruyère and Emmenthal notwithstanding.

Similarly, farming contributes no more than 2 percent to economic activity in the United Kingdom. In Ireland, RTÉ executives knew what they were doing when they ditched the ditches of *Glenroe* and stuck with the *Fair City* slickers. They broadcast to a nation where farming keeps 5 percent of the workforce active. It is greatly to the credit of all involved in agriculture that the sector chipped in with 9 percent of GDP at the dawn of the millennium.

For economists, food production is not something that twiddles their dials. Agriculture is not sexy. The Bangladeshis, for all the awesome fertility of their land, are down at the bottom of their graphs. The no-hopers in their tables are the countries most dependent on agriculture: Guinea-Bissau (pass the peanuts), Congo, Albania and Burundi. Top of the class are places where the population scarcely have a growbag between the lot of them: Hong Kong and Singapore.

Country	% Arable land	% Employment in agriculture	% GDP, agriculture	Leading export	GDP per head ($)
Bangladesh	75	63	30	clothing 73%	350
Ukraine	60	22	12	metals 39%	980
Denmark	60	5	4	manufactured goods 76%	33,040
India	57	60	27	textiles 24%	440
Hungary	53	8	5	machinery/transport equipment 52%	4,510
Poland	48	19	5	machinery and transport equipment 30%	3,910
Romania	44	39	23	textiles and footwear 34%	1,360
Czech Republic	43	5	5	machinery/transport equipment 41%	5,150
Italy	41	6	3	engineering products 35%	20,090
Thailand	40	49	11	computers and parts 14%	2,160
Spain	39	7	4	raw materials and intermediate products 44%	14,100
Bulgaria	39	14	21	base metals 19%	1,220
Turkey	36	43	18	clothing and textiles 41%	3,160
Nigeria	36	43	33	petroleum 91%	300
Portugal	35	13	4	consumer goods 41%	10,670
France	35	4	3	capital equipment 30%, farming/food 13%	24,210
Slovakia	34	9	5	machinery and transport equipment 37%	3,700
Germany	34	3	1	road vehicles 18%	26,570
Philippines	31	40	19	electrical and electronic equipment 58%	1,050
Pakistan	28	44	25	textile yarn and fabrics 26%, rice 7%	470
Netherlands	28	4	4	machinery and transport equipment 33%	24,780

Country	% Arable land	% Employment in agriculture	% GDP, agriculture	Leading export	GDP per head (
Greece	27	21	8	manufactured products 27%, food and drink 9%	11,74
UK	25	2	2	finished manufactured products 60%, food, drink, tobacco 6%	21,41
Taiwan	25	8	3	machinery and electrical equipment 50%	12,04
Belgium	25	3	1	vehicles 15%	25,38
Morocco	22	5	16	food, drink, tobacco 20%	1,24(
Vietnam	21	n/a	24	textiles and garments 14%, rice 12%	350
South Korea	21	12	5	electronic products 19%	8,60(
Israel	21	2	4	industrial goods 75%, agriculture 4%	16,18
USA	19	3	2	capital goods (excluding vehicles) 44%	29,24
Austria	18	7	1	machinery/transport equipment 42%	26,83
Indonesia	17	43	20	petroleum and products 8%	640
Slovenia	15	12	4	manufactures 46%	9,78(
Malaysia	15	26	12	electronics/electrical machinery 56%	3,67(
Cameroon	15	70	42	crude oil 42%	610
New Zealand	14	10	8	dairy, meat, fruit and veg 35%	14,60
Mexico	13	17	6	manufactured products 90%	3,84(
Ireland	13	5	5	machinery and transport equipment 37%, agricultural/foodstuffs 9%	18,71
Switzerland	12	5	3	machinery 29%	39,98
Japan	12	5	2	motor vehicles 15%	32,35

Country	% Arable land	% Employment in agriculture	% GDP, agriculture	Leading export	GDP per head ($)
Iraq	12	16	6	crude oil 100%	n/a
Cote d'Ivoire	12	60	32	cocoa beans, coffee 45%	700
South Africa	11	9	4	metals, metal products, gold, diamonds 65%	11,740
Iran	11	39	24	oil and gas 82%	21,410
China	10	50	18	machinery/transport equipment 27%	12,040
Argentina	10	8	7	agricultural products 56%	25,380
Singapore	8	0	0	machinery/equipment 63%	1,240
Russia	8	18	7	fuels and energy 40%	350
Kenya	8	80	26	tea 27%, coffee, horticultural 50%	8,600
Finland	8	6	4	metals, machinery, transport equipment 27%	16,180
Zimbabwe	7	68	19	tobacco 24%	29,240
Sweden	7	2	2	machinery (electricals) 36%	26,830
Hong Kong	7	0	0	clothing 40%	640
Brazil	6	26	9	transport equipment 15%	9,780
Australia	6	5	6	ores, minerals, coal, oil 46%	3,670
Venezuela	5	14	5	petroleum and products 70%	610
Colombia	5	19	13	petroleum and products 21%	14,600
Chile	5	15	9	industrial 45%	3,840
Canada	5	4	2	motor vehicles/parts 24%	18,710
Peru	3	7	13	gold, copper and zinc 37%	39,980
Norway	3	5	2	oil, gas and products 43%	32,350
Algeria	3	25	11	energy and products 81%	n/a
Saudi Arabia	2	5	7	crude oil and refined petroleum 88%	6,910
Egypt	2	34	18	petroleum and products 35%	1,290

Pocket World in Figures, *The Economist*, 2001 edition

32

THE GEE-GEES

> 'They shoot horses, don't they?'
>
> RACING CARS

'Plough an Irish acre and you walked eleven miles. It was a tedious job, one furrow at a time,' recalls Matt O'Toole. He was born in 1919 and is old enough to have followed a pair of horses up and down, up and down, from headland to headland, for a wage of maybe eight shillings a week. Eleven miles a day, six days a week – all for about 51 cent, plus four meals. His experience was that the employers who fed their workers well tended not to pay well, and vice versa: the good payers often skimped on the grub.

Matt reckons that he was aged about fourteen when he cycled into Enniscorthy town on his bicycle one day. He became aware of something extraordinary going on in a roadside field. Deacons of Monart were using a Fordson tractor to harrow their land. This was so unusual that a crowd gathered to watch, and admire the spectacle.

The man from Ballindaggin rates the invention of tractors as one of the great blessings enjoyed by mankind, but twenty more years elapsed before they became commonly used in Ireland. He continued plodding behind sturdy animals called Paddy and Billy or somesuch until he emigrated to the mechanised flat lands of Lincolnshire in England in 1948. He returned to buy his own Allis Chalmers tractor from Steacy's of Gorey in 1950, when he was to the fore in a gathering trend.

Official statistics suggest that in 1939 the twenty-six counties had 326,000 agricultural workhorses, ready to provide most of the muscle required to feed the country during the wartime dislocation of world markets. Not only did they plough, but they also harrowed, hauled out the dung and hauled in the crops.

In 1952, the trusty draught animals were still on call in large numbers, though the count had dropped slightly, to 286,000. During the following two decades, however, the horse was blasted away by the wholesale advent of the Fergie and the Ford. By 1975, fewer than four thousand showed up in the figures. This was mass redundancy. It is said that many of those laid off in the great cull ended up on French dinner plates, via a meat factory in the midlands. Others may have lined the stomachs of racing greyhounds.

'The horses are gone and they are no loss – they were heavy labour,' reckons Matt O'Toole, who is admirably unsentimental in his views. 'I never liked horses.' Yet he knew a time when almost every landholder with fifty acres had two horses and a pony – plus a donkey to keep them company, as like as not. Farmers with aspirations were likely to have a hunter too for point-to-point racing.

Even Matt has to acknowledge that the horses he worked with were intelligent. They responded to calls to go left or right. When they heard the dinner bell sounding out, they speeded up to finish a furrow, realising that a break was on the way. They were hardy and hard-working. They were everywhere. They were doomed.

Mogue Curtis is a stripling compared with Matt O'Toole. Nevertheless, born in 1936, he is still old enough to recollect a childhood when his late father's sturdy mare, known as Betty, produced a foal a year for farm work. In 1950, the family acquired grey tractor registration number MI 8327 from Larry O'Brien in New Ross for £405 (about €514). The price included a plough. Thereafter, Betty was serviced by a thoroughbred stallion in the hope that her progeny would make steeplechasers and hunters.

Mogue keeps the tradition alive with two piebald workhorses in his yard at Raheen. He competes in ploughing matches with Johnny and Paddy, both bred in Ballycullane. It is a hobby. He realises that the days are long gone when every parish had a harness maker, a blacksmith, a carpenter (to look after the carts and gigs) and a man who knew all the remedies for sick equines without having to call in the vet at enormous expense.

A class for working horses at the agricultural show in Killag, County Wexford in the summer of 2009 attracted just four entries: three from Wexford and one from Kilkenny.

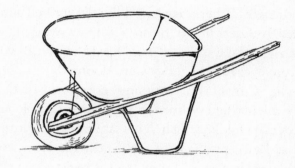

Thanks to Matt O'Toole of Ballindaggin and Mogue Curtis of Raheen

33

OFF-ROADERS

By no means everyone who takes part in ploughing events is a farmer. The fabulous Gill brothers from Ulster, Sammy and David, have both won world titles – yet they earn their crust as mechanics, not from the land. Here are a few more non-farmer ploughing fans:

GEORGE ACRES George describes himself as a smallholder these days, retired after thirty years' work as a haulier. He competes in the bright-red Farmall Cub that he won in a raffle at a ploughing match in Cavan a few years ago. He turned down the €5,000 cash alternative without a second thought. At the latest count, he reckoned that he had a baker's dozen of vintage tractors in his personal fleet, ranging in age from 1948 to 1969. They share his yard and sheds with a flock of poultry and several dogs.

JOHN WALSHE John works laying tarmac but at least he has eight acres at home to practise on. His competition tractor is a diesel-powered Ferguson 20 that was a heap of scrap when he bought it in Kilmyshall. A new engine has made the refurbished Fergie very nearly as good as it was in 1955. John proudly uses a Wexford-forged Pierce's plough, the last of which was manufactured in the sixties. He is a stalwart of the Shamrock Vintage Club, which, like many of its kind, raises large sums for good causes.

EAMONN WHITE Eamonn is a service engineer residing in Dublin. He bought the 1974 vintage Ursus in which he competes from a school in deepest Drumcondra. The school caretakers were using it for cutting the grass on the playing pitches. He concedes that the east European makes are not the sleekest-looking but insists that Ursus and Zetor once led the way in fitting advanced features such as compressors and air brakes for trailers. He grew up on a farm, and the decades spent in Dublin have not obscured his agricultural background. He returns home regularly: 'The country never left me.'

PADDY KENT Paddy is a company chairman, his name synonymous with steel products, but at weekends during the season he may be found aboard his ten-year-old John Deere, tugging a Kverneland plough. The interest in ploughing is a variation on a family theme. His late father had a small farm and competed in horse-ploughing matches, with considerable success, until he was into his eighties. Paddy's brother Jim tried his hand with the horses but it was Paddy who persevered, following a mechanised bent.

RONNIE COULTER Ronnie is the chairman of the Northern Ireland Vintage Ploughing Association, and one of the organisers of the 2011 European Vintage Ploughing Championships at Hillsborough. His day job involves knocking things down as a demolition contractor. His father was a very well-respected horse ploughman, but Ronnie's steed is a Ford 6640 with a reversible plough, rather than a vintage rig. 'Everyone had their own way of going mad,' he says cheerfully.

MICHAEL O'BRIEN Michael is a retired coach driver who has an alternative career breaking in horses, as well as a sideline providing a horse-drawn coach for weddings. He does not come of landowning stock but his late father worked for farmers throughout his life and the old man's duties included ploughing, of course. Now Michael follows in his father's footsteps, keeping a straight line behind heavyweights George and Trigger – one a big black-and-white piebald, the other a hefty animal of shire extraction. 'They eat a lot,' reports their owner.

Thanks to all of the above, of course

34

GRUB'S UP!

'I'm going out to dinner with a gorgeous singer'
CHRIS DE BURGH

'Come to the cookhouse door, boys'
TRADITIONAL MILITARY

When he sets out to plough, the experienced farmer, no matter how fastidious he may be, is not bothered about running his tractor through the car wash beforehand. The wise ploughman knows that, by the time the job is done, the vehicle will almost certainly be liberally covered in bird droppings. The chalky rain of guano is as much a part of the ritual as the smell of freshly exposed earth, or the packet of ham sandwiches on the dashboard.

PLOUGH MUSIC

The ploughing of a field is accepted as an invitation to dinner by hungry hordes of rooks and seagulls, and they respond in goodly number. The hapless invertebrate that finds itself suddenly exposed to sunlight by the mouldboard is a sitting target for the flock that follows the plough, and is likely to be devoured within seconds. Worms, leatherjackets and a host of other little morsels are brought to the surface – and they say there's no such thing as a free lunch!

Showing up Daz-white against the brown earth, the most obvious guest at this avian smorgasbord is the black-headed gull – *Larus ridibundus*. Meet Europe's number one scavenger, whose claim to be a gull of the sea is often no more than part-time. The black-head's range includes land-locked Austria and Switzerland, far from the briny.

While Irish black-heads prefer to nest along the coast, many of them are like well-to-do French families, migrating between city apartment and summer residence at the seaside. Ornithologists report a regular winter roosting colony at Gortdrum in Tipperary, just about as remote from the sea as it is possible to be on this island – and very handy when the call comes that the plough is out.

Ireland's only genuine full-time sea-going gull is the kittiwake. Other members of the family – including herring, common and black-backed gulls – are all likely to stray inland from time to time. Yet it is the black-headed variety that has forged the strongest links with arable agriculture. Granted, they have been joined traipsing after tractors in recent years by their cousin, the Mediterranean gull. However, the black-heads remain overwhelmingly the most frequent diner along the fresh-cut furrows.

The good clean taste of earthworm must make a refreshing change from hanging around landfill dumps and sewage farms, so they set about their picnicking with great enthusiasm and attention, in a feathery blizzard of constant noisy movement.

'They have no staple diet – they will eat anything. They are real scavengers, and pretty revolting if they regurgitate over you,' reports naturalist Chris Wilson, whose willingness to tangle with birds at close quarters clearly has its hazards. Moving on from mainly white plumage to black, Chris points out that the other principal member of the plough party is not the crow, as some observers lazily assume, but the rook.

Good old *Corvus frugilegus* is the farmer's friend, he says. Rooks may be raucous and rowdy, and life in the rookery may be characterised by inter-nest rivalries and petty larceny. But the rook is the best-organised bird in the business, and loves tidying up around the margins of agriculture. Unlike pigeons, they are not interested in eating corn, preferring to compete with the gulls for the protein-rich worms and other goodies uncovered by the plough.

Their table manners are much more refined than those of their sea-going colleagues. A flock of rooks may be seen picking through the newly turned sods with deep concentration. They resemble nothing so much as a bewildered party of wandering waiters in formal dress, sent in to tidy up after some monstrous garden party.

Grub's up!

Thanks to former wildlife ranger Chris Wilson

Collins Bird Guide by Stuart Keith and John Gooders, Chantacleer Press 1980

35

NOR PLOW

The World Ploughing Championships of 2007 (held in Lithuania) and 2008 (in Austria) attracted a combined total of 116 entries. The cosmopolitan mix of the events was reflected in the competitors' choice of tractors. Seventeen different makes were represented; Ford drivers won three of the four principal titles at stake, with New Holland completing the roll of high honour. The most popular choice of brand was Massey-Ferguson (26 of the 116), while Case (11), Valtra (10), Zetor (10) and New Holland (10) also featured strongly. One-offs were Landini, Valmet, Renault and the Swiss make Hurlimann.

The brand of plough used by the contestants, in contrast, was much more of a closed shop. Of the 116, only ten did not carry the green-arrow logo of Kverneland.

Andrew B. Mitchell from Scotland had the temerity to win the reversible section at Grafenegg in Austria pulling a Dowdeswell plough made in the UK. Other minority makes that appeared in the championships were Vogel & Noot, Rumpstad and Overum. Certainly, Kverneland is by no means the only major plough manufacturer – it just feels like that sometimes.

Take Kuhn, for example, who had a hand in the winning of the world title (reversible) in 2001 by Freddy Bohr. The company has serious pedigree. It was founded in 1828 by one Joseph Kuhn, a French blacksmith from near Saverne in Alsace. His enterprise specialised in weighing equipment before diversifying into agricultural machinery. Since World War Two, the company has followed the acquisitions trail, and they now have a substantial presence throughout the European Union, Russia and North America. But they are still only in the ha'penny place when it comes to competitive ploughing.

At the world event in Lithuania, Sweden's Anders Olssom turned up pulling an Oversum plough. Oversum's motto is 'Quality, Strength and Precision'. Anders was not strong enough, or precise enough, to do any better than trail in twenty-fourth among the thirty-one competitors in the reversible ploughing test. His compatriot in the conventional event, Johan Berggren, managed a respectable mid-table fourteenth place – with a Kverneland, of course. So much for flying the flag.

The record of the Steeple Bumstead Agricultural Discussion Society meeting on 14 October 1997 proclaimed proudly: 'Dowdeswell are a British success story.' The story originated in the requirement of Warwickshire farmer Roger Dowdeswell for a compact plough. His home-made design proved so popular that he had to set up a factory to cope with demand. A quarter of a century after it was established, Dowdeswell Engineering could claim to be the largest British manufacturer of agricultural implements. Still, small beer, all the same.

The English plough maker is a boutique in comparison with the Austrian giant founded in Wartberg during 1872 by Friedrich Vogel and Hugo Noot. They made shovels. Ploughs did not feature until more than twenty years later. In 1939, Vogel & Noot ran off seven thousand of them for pulling by horses. By 1994, they could plausibly claim to be the second-largest plough manufacturer in the world. But not as big as you-know-who . . .

Kverneland is named in honour of Ole Gabriel Kverneland. In 1879, he opened a forge to make scythes in the village of Kvernaland near Stavanger in Norway. The firm made its first tractor plough in 1928. It remained in family hands until 1983, when the company was floated on the Oslo Stock Exchange. By then, they had already snapped up Plovfabrikken Fraugde of Denmark, in 1973. In due course, other purchases were added: Kylingstad Plogfabrikk of Norway (1984), Underhaugs Fabrikk of Norway (1986), Maskinfabriken Taarup of Denmark (1993), Machine Agricole Maletti of Italy (1995), Accord

Landmaschinen of Germany (1996) and Greenland Group of the Netherlands (1998).

In 2000, they spread to the Loire valley, and absorbed French company Gregoire-Besson, though not so much because of their plough-making tradition. The move was made in order to effect entry to the vineyard. The aquavit-drinking Norwegians are now the biggest thing in viticultural machinery as well as ploughs, though the two strands are operated as separate businesses. With factories in Denmark, Germany, France, the Netherlands, Italy and Russia, they have come a long way since Ole Gabriel used a water-powered hammer to 'mass-produce' scythes – at a rate of seven thousand a year.

www.worldploughing.org

www.gregoire-besson.com/gregoire-besson-histoire.php?lg=en&id_rub=1&page=experience#

www.sbads.co.uk/news05.html

www.overums-bruk.se/overume/engbest/xcelsior_eng.pdf

www.vogel-noot.info/index.php/article/view/126

www.kvernelandgroup.com/welcome

www.kuhn.fr/fr/gamme-labour.html

36

A Plant in the Wrong Place

'Picking up lots of forget-me-nots'
SUNG BY FRANK SINATRA (J. MYROW/M. GORDON)

The Hundred Years War lasted more than a century, but the struggle by farmers with the forces of hempnettle, knotgrass and nipplewort is a perennial guerrilla struggle on a million fronts that shows no sign of ever ending.

The enemy may have cute names such as corn marigold or parsley piert. The enemy may have sinister names such as blackshade or red deadnettle. The enemy, such as forget-me-not, may be found occasionally in respectable places. But never let us forget that the enemy is still the enemy.

PLOUGH MUSIC

One of the big selling points of ploughing is that it buries last season's weeds out of harm's way. One of the great disadvantages is that the plough turns up weed seeds from previous years and liberates them, to wreak their havoc. The enemy is capable of cutting grain yields by a third and cannot be permitted to stand unchallenged in the field.

The war has long since entered its weapons-of-mass-destruction phase, with a range of lethal chemicals to hand for battering the enemy into temporary submission, though never outright surrender.

Many of the herbicides come with suitably sinister labels – Thor, Finish, Hussar, Tomahawk – like characters in some violent computer game from the *Mad Max* stable. Others have calmly reassuring titles – Pacifica, Harmony, Ally. Some assume an impressive scientific guise – Oxytril, Fluxyr, GEX, Stellox. None of them has a perfect strike rate. Their targets build up resistance and bid the laboratory teams do their worst.

The successful grower needs to be able to spot the intruders when they are just poking their dainty little heads above the dirt. By the time scarlet pimpernel has sprouted its scarlet flowers, the moment for decisive action is long since past. By the time sowthistle has sprouted its prickles, it is a problem already beyond a ready solution. Once allowed to grow rampant, the enemy is capable of robbing the soil of the goodness that makes for a good crop. The enemy can block essential light or pull down otherwise healthy stalks of good grain.

The enemy is a moving target that comes in many forms. Here are some of the worst:

CHICKWEED Here's one chick that no self-respecting arable man wants to be seen with. She is rated 'troublesome and widespread' by the weed experts. Also known as chickenweed, winterweed or white bird's-eye.

REDSHANK has more nicknames than the Bunclody junior hurling team, among them Christ's-a-bleeding and arse-smart. The allusions to 'red' and 'bleeding' are inspired by the weed's crimson stem. Like the hurlers, redshank has no objection to heavy ground.

BLACK BINDWEED Sometimes referred to as climbing buckwheat, it is a deep-rooted menace that seriously affects corn yields and snags combine harvesters. Its first true leaf is heart-shaped – but don't be led down lover's lane by this pesky valentine.

FUMITORY Called beggary in some places, it is a curse, with its long-lasting seed and its ability to resist most of the herbicides sent to destroy it. The name God's-fingers-and-thumbs is a reference to its leaves, though they look more like the hands of a newborn infant than of any deity.

FAT HEN grows to be fat on the nitrogen that would otherwise be nourishing the legitimate crop among which it flourishes. Also goes by the names of dungweed, goose foot, lamb's quarters and muck-weed.

CLEAVERS The Latin moniker *Galium aparine* gives a cloak of respectability to a grade-one menace. Cleavers has seen off many of the sprays sent to expel it from tilled ground. It continues to pull down stems of grain and catch in reaping gear, and its seeds are a problem in amongst good honest cereal. It has accumulated more aliases than any public enemy number one: ariff, goosegrass, herriff, sticky-willy, robin-run-the-hedge, beggar's lice, scratch grass, cliver.

And that's just some of the broadleaved pests. The grass weeds are another bunch of bandits. There's the Bent Gang: black bent, creeping bent and loose silky bent. There's Brome Boys and the Fiercesome Fescues. The list runs to twenty-one, including the seductively titled awned canary grass.

The fight continues.

'The FBC Guide to Grass and Broad-Leaved Weed Identification', 1984

Thanks to Phelim McDonald, Teagasc, Enniscorthy

37

THE WEST AWAKE

'the green, green grass of home'
SUNG BY TOM JONES

No one will ever make their fortune selling ploughs in the west of Ireland, that's for sure. The south-east and south-west regions boast 46 percent of the Republic's specialist tillage farms. They have the free-draining soils and comparatively warm climate needed for the job. Ocean breezes are not conducive to high yields of wheat or barley, it seems.

The national ploughing championships are a truly national event. However, for all the razzmatazz that surrounds the annual jamboree, arable farming is, and always has been, a minority interest across Ireland. In 1991, cereals were grown on just over 23,000 farms in the Republic – less than one in every seven out of the total of 178,000. Typically, the holdings with the tillage were on the sweep from Cork around to Louth.

The trend towards concentrating tillage in specialist enterprises has not slowed since. There are few grain-growing hot spots along the seaboard from Kerry, through Limerick, Clare, west Galway, Mayo and Sligo to Leitrim. Donegal, in the shelter of the Finn valley, is one county that bucks the trend, along with sheltered patches of east and south Galway.

Even potato growing, at least on a commercial scale, is not the preserve of the hardy west-of-Ireland grower. At the beginning of the nineties, 12,400 farms were recorded as having spuds, but potato power was (and remains) concentrated in the hands of a few. One-tenth of the growers produced 63 percent of the crop. The big lads are located in Cork, Kilkenny, Wexford, Dublin, Meath and Louth. Donegal is the only western outpost of potato power, as its cool clime is ideal for seed spuds.

Just because there is so little arable activity along the Atlantic seaboard does not imply by any means that there is no ploughing. For instance, the Corbett family from Clare has produced All Ireland-title-winning ploughers (if there is such a word) in three successive generations: the late John Corbett started the sequence in the fifties, to be succeeded in turn by his son Fran and then by Fran's daughter Lorraine.

Gerard Frost junior from Kilkishen in the parish of O'Callaghan's Mills in Clare has also mounted the rostrum at national ploughing championships. Gerard works for a machinery company and lives in a part of the world better known for bare limestone rock than for its cornfields.

'You would be very lucky to see a ploughed field in Clare,' admits county ploughing association official Noel Fitzgerald. 'We should nearly get a prize for going to the All-Ireland, let alone competing. The only way we have a ploughing match is where a farmer is re-seeding lea ground.'

The tradition of grain growing around Newmarket-on-Fergus has largely petered out, but an interest in vintage farm equipment is the mainstay in keeping the tradition alive, not only in Newmarket but also in Sixmilebridge, Cratloe, O'Callaghan's Mills and Parteen, with occasional outbreaks of interest elsewhere across The Banner.

Anthony Brennan of the ploughing association in Leitrim has to concede that his county is more renowned for its rushes than its fields of waving corn. Nevertheless, he makes no bones about claiming the annual ploughing match at Carrigallen each Easter weekend as the biggest of its kind in the west of Ireland. The event produces champions who go on to rival the best at the national ploughing championships. With no stubble available, competitions are staged on good green grass, which has to be re-seeded and repaired afterwards by Liam and his committee. The tradition of horse ploughing survives strongly, with little enthusiasm for the high-tech reversible ploughs.

> In 1933, Irish farmers were given an incentive to grow tobacco, when duty of nine shillings (about 57 cent) was removed. County Wexford farmers grew tobacco over the period from 1904 into the forties, though the quality of the crop was variable. Farmers who fancied themselves as players (ahem) in this market were urged to plough farmyard manure into the chosen plot of land as early as possible in the winter beforehand, and not to stint on the sulphate of potash.

Irish Agriculture in Transition by Séamus Lafferty, Patrick Commins and James A. Walsh, Teagasc, 1999

38

Mud, Mud, Glorious Mud

'Come let us wallow in glorious mud'
                      Flanders and Swann

'One could meet and converse at ease and relaxation with a farmer from Skibbereen, Haverford West, Kenya or Ohio. Good humour and laughter was the common denominator shared between the men of land from all over the Globe.'

So reported Michael T. Connolly as he reviewed the staging of the World Ploughing Championships at Rosegarland, County Wexford, in 1973. They were run in tandem with the national championships.

The good humour adverted to by the chief agricultural officer was very much required. The weather turned nasty on the Friday, making wellingtons an absolute necessity. This was in the days before the organisers invested in the metal walkways that are nowadays laid across the chosen fields to offer some semblance of firm footing. Back in 1973, the hundred thousand or so spectators made their own arrangements.

The gales caused the temporary collapse of the marquee which had been erected by the Clongeen tug-o'-war club to run fund-raising dances during the course of the championships. The club needed the dosh for an expedition to Europe and they had a lucrative hit act on their hands in the genial form of Big Tom (with his Mainliners). The big tent mercifully stayed erect on the night the band thundered through their repertoire. No alcoholic drink was served under canvas but the gate receipts were enormous and Jimmy Curtis recalls club members sitting comfortably on bagfuls of money that night.

His son Frank was just nine at the time and he retains memories of the excitement in the neighbourhood during the countdown to the championships. Schools in the area were circulated with a poster bearing the flags of all the competing nations, of which there were at least nineteen. The tug-o'-war club was not the only local organisation to seize the opportunity to make money, as the GAA, Macra na Feirme, the IFA and other groups produced teams of volunteers for stewarding and to supervise parking.

The visiting champions from around the world stayed in Wexford town and were bussed to and fro between their hotel and the 400-acre site on the land of the Leigh family. Still there was still plenty of demand for bed and breakfast from lesser mortals and there was scarcely a bed to be had in the greater Clongeen area for the duration.

The Curtis family provided accommodation of a different kind, allowing competitors from Northern Ireland to use their farm as an unofficial practice ground. Such diligence was commendable, but the Ulstermen finished off the pace as Scandinavian contestants dominated the event, along with Carlow's John Tracey, who finished runner-up.

For those who did not fancy scrumming down with Big Tom, the official music programme offered the Artane Boys' Band and the Army Band. Other entertainments included sheepdog trials, clay-pigeon shooting and demonstrations by the Irish Countrywomen's Association. All this was in addition to the 120 trade stands – and don't forget the ploughing, of course, though many among the multitude scarcely gave it a glance.

Rosegarland, it may be remarked, is a lunatic place to hold a crowd-pulling international event. Fifteen miles from the nearest town. Served by a network of back roads. When the national ploughing association hit upon such an obscure venue, however, they were sticking to their custom of keeping it country, to coin a phrase. They were still smarting from the experience of two years previously, when they had gone to Finglas for the national championships. Their target audience was apparently daunted by the prospect of tangling with the Dublin traffic, and the association recorded a loss in 1971. Navigating the intricate byroads of south Wexford held no such terrors for the followers of the plough.

Ireland has hosted the World Ploughing Championships on nine occasions. Killarney in Kerry (1954), Armoy in Antrim (1959), various farms at Limavady in Derry (1979, 1991 and 2004), Oak Park in Carlow (1996), and Tullow in Carlow (2006) are all on the list. Only Rosegarland appears twice. The jamboree, first experienced in 1973, returned with President Patrick Hillery among the guests in 1981. The rain turned conditions underfoot to the consistency of brown pancake batter.

And no one minded much at all.

Thanks to Jimmy and Frank Curtis
http://homepage.eircom.net/~nationalploughing/history/1973.html
Michael T. Connolly, *op. cit.*

39

THEY DIDN'T LICK IT OFF THE STONES

'Here we go, rockin' all over the world'
STATUS QUO

People have been ploughing for many centuries but no one has acquired more kudos as a master of the art than Martin Kehoe, from Clongeen in County Wexford. Three times world champion, his record is matched only by Northern Ireland's Hugh Barr (a fellow tug-o'-war enthusiast, by the way), who took the crown in three successive years.

The first thing you notice about Martin is his hands – hands as big as shovels. Born in 1949, he is the original gentle giant, and one-time holder of the title 'Ireland's strongest man'. Hauling on a rope has earned him plenty of medals, making obvious use of his physique. Yet it is the cool brain atop the brawn that has propelled this hard-working farmer to superstardom.

His late father, Willie, was the inspiration, winning five national crowns, starting with horses and then moving on to mechanised means. Willie was the product of an era that was anything but mass-produced. His 'Star' brand ploughs were custom-built by Wexford Engineering to meet individual requirements and then further modified at home. Martin still remembers mouldboards being sculpted by his dad on the hob in the family kitchen.

The son was smitten early with the bug. He used to come home from national school and hop on to the tractor to make his mark on the fields of Clongeen. As a teenager, he was soon catching the eye further afield. He won his first ploughing match, at Bree, in 1965, going on to represent his county at the national championships in Enniskerry, where he won the under-21 grade at the first attempt.

'One place I don't like going is towns,' says the oft-times champion, but his distaste for urban life has not prevented him from developing a truly international perspective on life. He repeatedly represented Ireland abroad, his reward for complete domination of the scene at home for a dozen years from 1988. Christmas cards arrive in Clongeen each year from Holland, Australia and many places in between. His three world titles came on three different continents.

Martin's first appearance on such an exalted stage was in 1982, in Tasmania. The Tasmaniacs were much impressed by his Lamborghini tractor, which came with a Ransome plough, but his performance did not match that of the racy Italian marque: 'The soil just stuck to the boards', he said, and he was left muttering about marl holes.

He finally made the breakthrough to the top of the podium twelve years later. A story is attached to each victory on a foreign field. In 1994, at Outram in New Zealand, he was fortunate to be assigned a pitch on good ground which only required ploughing to a depth of six or seven inches. The first-time winner left South Island imbued with a great fondness for the country.

In 1995, when the ploughmen of the world gathered on the peaty black soil of Njoro in Kenya, Kverneland provided plastic mouldboards. Power supply was erratic, so when the Irishman decided on some rudimentary alterations, he had to set about the re-designing work with a hacksaw. His father would have been proud of him. Some of the credit must also go to his local dealers Duncormick Tractors, who arranged to have a Landini waiting for him in far-off Njoro.

In 1999, at Pomacle near Reims in France, the contest was memorable for the hordes of mice that invaded the field. The little creatures popped up in swarms from the grassland, and many were crushed under the big rubber tyres during the two days.

The essential ingredients for a winner include an eye

for a straight line, a readiness to react to whatever the land and the weather throw at you, attention to detail, and a good pair of spanners. A taste for Irish country music is optional.

The only other winner of the world title to come out of the Republic was Charlie Keegan of Wicklow (who won in 1964). Those who have attempted to scale such heights over the years have included a remarkable number of individuals from the Boley/Clongeen area of the Model County: Andrew Cullen, Richard Byrne and Frank Curtis, all from the valley of the Owenduff river. Michael Keating of New Ross was originally from Boley too, and his son Seán has also made the grade. Also on the list is one Willie John Kehoe, son of Martin. Other siblings, Eleanor, Christina, Michelle and Martin junior, also compete.

They didn't lick it off the stones.

Thanks to Martin Kehoe

40

How to Make a Small Fortune from Farming? Start with a Large One

'Is it worth it?'
　　　　　　　Elvis Costello ('Shipbuilding')

The great tradition of agriculture is not necessarily also a great tradition of money-making. According to the data in the annual Teagasc farm survey, the Celtic Tiger bypassed farming completely. The average farm generated just over €14,000 in 1995, and this rose very close to €17,000 by 2008. Yes, the trend was up. But the clever number-crunchers at the Central Statistics Office make allowance for inflation and point out that this is in fact a reduction in real income to the tune of 22 miserable percent.

Of course, most grain producers would be vexed to have themselves described as 'average'. They tend to be at the biggest and best-organised end of the farming spectrum. Nevertheless, they must worry at the way that their figures do not add up to economic sense. It is small wonder that the average age of farmers is increasing as the younger generation takes a look at the dwindling returns and says 'no thank you'.

Is it worth it? Not if you aspire to steady earnings. The income stream from farming is as dependable as gambling – without the fun of going to the races. The figures show that tillage-farm income drooped 5 percent in 2006. The following year, income per farm went up an impressive 41 percent – only to be followed by a sickening 52 percent cut as the roller-coaster ride took a stomach-churning lurch downwards.

Is it worth it? Many must be having doubts. According to the *Irish Times* in January 2009, a newly qualified teacher's basic salary is an annual €32,599. Compare that with the average tillage farm, where the income in 2008 was less than €20,000, with little prospect of any improvement on the horizon for 2009. Meanwhile, the teacher's pay is set to rise, through measured increments, to €65,807. At least the grain grower is ahead of the freshly enlisted square-bashing army private, who draws €13,403. The Teagasc survey reported: 'Overall, on 79 percent of farms, the farmer and/or spouse had some source of off-farm income, be it from employment, pension or social assistance.' Given the increasingly meagre returns from the land, it surely makes sense to have something to fall back on: for many, agriculture must be the sideline.

Is it worth it? In spite of all the above, those involved still appear to believe that it is worth taking a punt on there being a future for their industry. Tillage farmers continued pumping money into their enterprises, at the rate of €18,000 per farm during 2008. It seems that the prospect of the next bumper crop is as compelling to the farmer as a maiden handicap at Punchestown to a compulsive follower of the gee-gees: 'Gross on-farm investment in buildings, machinery and other assets was estimated at just over €2 billion in 2008 – the highest annual level of investment ever recorded on Irish farms,' says the author of the farm-income survey.

How the hell can it be worth it?

41

THE SECRETS OF THE ART

So, how is it done? How the hell do you actually plough? No better man to put the process into words than Eamonn Tracey of Garryhill in County Carlow, who won the first of his senior national ploughing titles in the year 2000. Eamonn is the son of John Tracey, who won the first of his seven (and counting) titles back in 1973, and has been runner-up in world championships five times. They plough forty acres each year on the home farm and then carry out some contract work too.

'Basically, it's not how you plough, it's how you set up,' Eamonn says. 'You buy an implement; you have the machine to do the work. How well a field is ploughed depends on how well the operator sets up the plough.

'Our plough here is a four-furrow reversible Kverneland. The Kverneland is the most popular plough in Ireland. It is probably the best for Irish conditions. Other manufacturers mightn't like to hear me saying that, though the others are coming around to making mouldboards for the Irish market. The mouldboard dictates the way the soil is turned.'

What preparation is involved? 'With regulations as they are now, you can't plough for twelve weeks between October and January. A good day, maybe two, would get your plough ready for a season to start off. A lot of replacing of wearing parts is involved. The plough consists of many wearing parts.

'In the order they hit the ground, you have the disc, the circular things that cuts the sod. The skimmer runs beside the disc. The soil is skimmed before, or at the same time as, it is lifted. Then you have the point and the share. Then you may have a shin piece, on the front of the mouldboard, and that is the most-wearing part of all.

'Mouldboards wear out too if they hit too many rocks. Irish soil is heavy-wearing soil. Nearly all fields in Ireland have little stones, grit, in them. You take a plough, ploughing all day — it's constant abrasive motion against it. Anything that touches the soil wears out. For replacements, Kilkenny has the main Kverneland dealer, or there's a place, Heggarty's of Tullow, or you can buy all spurious parts made to suit the Kverneland.

'The pre-season work is about making sure everything is going right.'

How deep down do you go? 'It depends on the depth of soil in an area. Around here in Garryhill, there is no great depth of soil. The maximum you could plough here would be, say, seven inches. You could plough eight inches here but then you are only digging up subsoil, and that's no good. Once you head down to Fenagh, and better ground, you would be going nine inches if you wanted.

'Ploughing for a root crop, you would always go as deep as you could, to allow the roots to penetrate the soil. For cereals, you don't need to go as deep, because their root system is at the top of the ground. But then you always want to plough deep enough.

'Our plough is fixed at a width of fourteen inches, but the trend now is towards vari-width ploughs. Each furrow of ours is fourteen inches wide. But the vari-width ploughs can be varied from twelve to twenty inches by using the hydraulics on the tractor.

'The temptation is to set it for twenty inches but, if you are ploughing twenty inches wide, by right you would probably want to go deep to get the sod to turn properly. And that is the problem with a lot of contractors. The deeper you go, the more wear, because you're down in harder ground. They want to be in and out as quickly as possible, so what they do is they widen out to the twenty inches and still go down just seven. A lot of bad ploughing is done as a result. They don't bury a lot of the trash properly, and the whole thing is left too loose.'

What are the pitfalls? 'You can damage a field by ploughing it badly. You can leave humps and leave the ground terribly un-level. I've seen a field where a lad went into a grass field. He widened to twenty inches and left eight of those inches practically undisturbed. That was a mess. We had to go up with a Rotavator and chop the whole thing up.

'You see a lot of bad ploughing work at headlands. There's an art in getting that done right. You often see fields left very un-level. It's all about marking a field out level. Even in commercial ploughing, there is a bit of planning involved. An awful lot of contractors just throw a young lad up on top of a tractor and send him on.

'Modern tractors compact the land more than the old Fergie 20 used to. It takes management to avoid compaction. There is no use going in on a field when it's just after spilling rain – you're doing harm. You have to wait until it's dry enough or you can do a lot of damage. If you turn over a field that is saturated wet, you will never dry that. When you then come in to sow, you are really compacting it.'

Is it worthwhile work? 'When I go out the gate after ploughing a field, it's not the same as competition, but I take satisfaction from doing the job all the same. When we plough, we still try to keep it straight and even. If the furrows are not even, it has an effect on the operations that follow. You go to spray where the ground is not level, for instance, and the booms of your sprayer are bouncing up and down. Or if you are spreading manure and your tractor is hopping, then your spreader is waving up and down. When it's down, it's only spraying out four yards, but when it's up, it's spraying twelve yards. It's important for tillage to have your ground level.'

Any tips? 'Prepare a field beforehand. If you have a field that's weeds and scutch grass, the best of ploughmen won't do a good job on that. Better apply herbicide. That's another cause of bad ploughing – going into fields that are not suitable. Scutch grass especially can cause horrendous problems.'

What's the difference between competitive and commercial ploughing? 'I ploughed at world championships in France and Germany and Lithuania. I ploughed in Northern Ireland and I ploughed just over the road at Rathoe in 2006. It's hard to beat the Irish soil. Irish soil is the best you'd see, but the weather here is so wet. If we could get European weather! Though even as it is, we still break all records. The farmers here are growing better crops than anyone in Europe, where they have decent growing and harvesting weather – and they are still not able to produce the tonnage per hectare we are getting, particularly in wheat.

'Competition ploughing is a sport, and very little to do with looking after a field. We still compete in two-furrow conventional ploughing. It's a style that is just not practical commercially. You are really doing it as an art or a sport. It has nothing to do with what you do on the farm at home. The way I have my match plough set up, it would plough a lot better than the plough we have for commercial work.

'But a good day's commerial ploughing is fifteen to twenty acres. A long day would plough a twenty-acre field with the four-furrow reversible set-up. But in competition at the world championships, you would have two hours and fifty minutes to plough a plot one hundred metres by twenty metres. That would be two thousand square metres – a fifth of a hectare, exactly half an acre. In a ten-hour day at that rate, you'd only have five acres done. Not practical.'

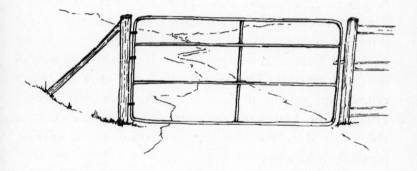

Thanks to Eamonn Tracey and his father John, many-times national ploughing champions

42

EMERGENCY MEASURES

'Promised you a miracle'

SIMPLE MINDS

'Farmers have done miraculous things,' said Offaly farmer Roy Fossitt, who was born 1934. 'Make it worthwhile, and they will come through with the wherewithal.' The observation was made as conversation turned to the years of the Emergency, when most of the rest of the world was at war and Ireland had no choice but to fend for itself in lonely neutrality. It was time for Irish food producers to deliver their miracle.

'One more cow; one more sow; one more acre under the plough.' Ask any country person of Roy's generation – he was a young boy in the Emergency years of 1939–45 –

and they will parrot back the rhyme. These were the days of the compulsory tillage orders, when the nation expected that every landholder would do his (or her) duty. Some did so with greater enthusiasm than others.

In the case of the Fossitts, Roy's father – the late Harry Fossitt – reacted badly to the word 'compulsory'. He hated being told what to do. So he declined to install logs in his field, as required, to discourage German aeroplanes from using the cattle pastures of Walsh Island as the entry point for their feared invasion.

Still, Roy reckons that the da did actually plant an extra acre or two of wheat, though he never boasted about it. And he fed some of the precious crop to the pigs, which was strictly against the rules. As the owner of a tractor, he made his machine available to plough the fields of neighbours. Of course, he always maintained that this was less out of patriotic ardour than the desire to obtain the fuel coupons needed to keep the precious Ford in action.

Harry was not the only one dragged into the wartime grain game against his natural inclinations. The farmers of the west groused at having to drop their sheep numbers to make way for the unfamiliar business of grain cultivation. Many Meath cattle specialists were sent into a tizzy as there was scarcely a harrow, let alone a plough, in entire parishes – though the soil there was lush, and perfectly suitable for the job.

On the other hand, there were many eager to heed the call and grow their 'one more acre' – then one more acre again, and yet again. At the Kings' family farm in County

Kildare, the rhyme was ascribed to Éamon de Valera. As a loyal member of Fianna Fáil, Tom King did not hesitate to put his 133 acres at the disposal of the cause. Tom dutifully put down his logs to stave off the Luftwaffe.

He, too, had a tractor – one of the last to come into the country before the outbreak of hostilities put an end to such imports. The yellow-painted Dagenham-built Fordson was supplied by dealer Charlie Taylor in Athy with cleats on the wheels rather than expensive rubber tyres. It came with a three-sod Ransome plough, which was worked hard for the duration of the Emergency. A great deal of nervous energy was expended finding the supplies of petrol and TVO fuel required for the tractor.

Seán King, who was born in 1933, still recalls the excitement generated by the arrival of the yellow machine at Rathgrumley. In fact, this was the machine on which he learnt to drive. It was a thrill for the young man to assist his father in tugging furze bushes from the ground in the farm's pasture field to make way for the grain that would keep Ireland decently fed.

'Let her on!' Tom would cry when the chain and hook was in place on a bush. Then Seán would gun the accelerator and duly let her on. Their efforts at clearing the land uncovered reminders of a previous emergency. They discovered the undulations of lazy beds where Famine-time folk grew potatoes in the 1840s.

Not all the results of the 'one more acre' initiative were top quality. Seán's wife Mary recalls that a loaf of allegedly white bread bought from the baker in Athy often had a black streak. More often than not, farm families avoided the full rigours of the rationing regime and the dodgy shop-bought bread.

Tom King grew so much corn on his lime-rich land that it took five days to thresh the corn. No one objected if a little of the harvest never reached the market. Seán's mother diverted some of it to relations at Nurramore who had a mill. The flour was later sieved through the fine mesh of a lady's stocking to remove the husks and kept in a big bin near the fire in Rathgrumley, ready for baking.

Thanks to Roy and Lila Fossitt, Seán and Mary King, and Adrian King

43

From Athlone to Palataka

'They poured across the border'
           Sung by Leonard Cohen

John Garry was raised on a small farm in Athlone. The family holding has since disappeared under urban sprawl, with factories and a Lidl supermarket doing business where his father once milked cows and grew potatoes. As a member of the St Patrick's Missionary Society, John has spent most of his adult life far from the midlands, in East Africa, where he has participated in agriculture of a very different kind from that now practised in his native land.

John was ordained a Catholic priest in 1969 and was dispatched to Kenya, a country with some marvellously fertile and productive areas – as well as vast expanses of near-desert. His billet was a place called Baraka, where he arrived to find farming practice lagging far behind what would be considered modern back home around Athlone. Certainly tractors were completely unobtainable, unaffordable and unheard of at the time – at least in the parts of the country away from the commercial enterprises of the Rift Valley.

Yes, the people raised corn, but the maize they grew was tended to by hand, from preparation of the seedbed through to harvest. The principal instrument of husbandry was the hoe – the *jembe*, in local parlance. The man from the dairy farm in Athlone was expected to show leadership on the land, as his diocese had bought a demonstration farm. He was able to show how choosing seed carefully could improve yields, but the work was still carried out by hand. Subsistence was the name of the game.

While in Kenya, he eventually coaxed farmers into taking a step forward and using cattle for ploughing – flying in the face of the commonly held view that cows were not for work but for admiring. Later, with the assistance of Gorta, he was responsible for bringing a second-hand Massey-Ferguson 125, trailing a two-disc plough, to Baraka. Many things have changed in Kenya.

John's colleague Tom McDonnell has the sight of horses ploughing the family's holding in Antrim among his boyhood memories. He was impressed by the cooperative spirit demonstrated by the folk he worked with in the

Katali Highlands of Kenya in the mid-1970s. Fifty families at Matunda clubbed together to purchase a Massey Ferguson 135, and a second tractor was added in the eighties.

In 1983, John Garry was sent over the border into Sudan, to the Torit region, where Christianity was the common religion but where the Islamic government viewed missionaries with suspicion. Once again, farm work was carried out by dint of hard physical labour – at least until a party of Norwegians arrived to promote the use of oxen for ploughing.

The parish of Palataka in Sudan had a tradition of growing sorghum. Mention sorghum, and John says 'birdseed' – it's a reflex that comes with a mirth-free laugh. The crop often attracted hungry plagues of avian raiders capable of cleaning out his parishioners in a matter of hours. Maize was a more palatable (and more bird-resistant) alternative, though it was vulnerable in drought conditions. Coffee was encouraged by officialdom as a cash crop.

Father John had the run of six square miles to farm in this remote district, but he reckons that only eighty acres were actually viable farmland. In the end, it was neither the lack of rain, nor the marauding birds, nor the collapse of the coffee price that forced the Irishman and his flock to flee from their land in Palataka. The war that rumbled around Sudan for more than a generation finally caught up with the neighbourhood in 1993, and they sought shelter in a Ugandan refugee camp.

Tom and John both worked at the camp in Uganda,

where everyone helped everyone else to make the best of strips of land that were little bigger than allotments. However, while some areas of East Africa have made progress, those refugees who returned to Sudan have largely failed to establish firm economic foundations.

In a world where some farmers have combine harvesters as wide as commercial jets and tractors can be operated by remote control via satellite, the people around Palataka have almost given up on agriculture altogether. They would prefer to depend on international relief supplies than risk their efforts being looted by the various military and militia groups that still infest much of Sudan.

With thanks to John Garry and Tom McDonnell

44

THE THIN BLUE LINE

Ireland does not host G8 conferences, with all their attendant pomp and unrest. When U2 perform concerts in their home country, the venues tend to be in major cities. So a Garda stationed in a provincial area can scarcely be handed responsibility for a fixture of higher profile and complexity than 'the ploughing match'. The national championships annually attract spectators in the tens of thousands, creating a good-natured tsunami across the normally still pond of some rural backwater. In terms of police logistics, assignments seldom come bigger than the ploughing championships.

Superintendent Peter Finn has had the privilege of handling two such gatherings. The quietly spoken Kerryman has nothing but fond memories of his experiences in 1994 (when the championships were held in Drumgoold, Enniscorthy) and 1998 (Ballycarney, Ferns). He was first handed 'the big one' when working as an inspector in Gorey. And he made such a good job of it that he was back for more four years later. His superiors clearly liked his approach to major events: he also took charge of golf's Irish Open on one occasion.

As a countryman, Peter approached the ploughing-match assignment with a sympathetic, empathetic, accommodating attitude. In his youth, the superintendent had himself ploughed with a pair of horses – one black and one grey – on the Finn family's thirty-five-acre holding in the west of the Kingdom. The frost-free Kerry climate was considered ideal for growing early potatoes, and the farm also produced onions for the local St Brendan's co-operative. His first experience of ploughing as a sport, however, came in 1973.

Then aged twenty-one, he was a raw Garda recruit when he was sent from the barracks in Arklow to rain-swept south Wexford. That year's Rosegarland championships are still remembered for putting the wellington into Wellingtonbridge. He carried out his duties in a sea of mud, sustained by official rations that extended to a bottle of Taylor Keith and a ham sandwich. By the time he had risen through the ranks to put his own stamp on proceedings, at least the officers under his command could expect better fare.

Setting up a decent Garda catering unit was only one small part of an exercise that took the best part of a year to plan. Peter Finn was recruited because he was working in the traffic corps at the time. This makes sense, as traffic management is one of the principal responsibilities handed to the force come the ploughing match. In the case of Drumgoold, the task of coming up with a plan to bring in all the thousands of visiting cars smoothly to Kavanaghs' farm was complicated by the presence of Enniscorthy town, less than a mile down the road.

A one-way system was imposed on the motoring public, with the added convolution that the one-way of the morning became the other-way in the evening. An exclusion zone kept vehicles that were not bound for the main event at bay, but the ring of steel had to accommodate hundreds of local drivers, who were issued with large, bright pink resident passes. On the first morning, all appeared to be going well – too well. News came through that the Munster hordes were being stalled at the bridge in New Ross, more than thirty kilometres away. Motorcycling reinforcement was sent pronto to Ross to tackle the tailback.

The cop on the bike was one of 120 Gardaí assigned to the championships in both years, working in two shifts of sixty. Manpower was required as soon as the first piece of machinery was brought in by the first exhibitor: the force was in charge of site security. In the days before mobile ATMs, protection had to be provided for a temporary bank branch too. Officers, with back-up from Gerry O'Neill's communications centre, were detailed to look after visiting dignitaries, including Cabinet ministers.

Though they came armed with the full powers of the Public Order Act, the Guards were more likely to find themselves dealing with lost children than misbehaving drunks. The late (and genuinely great) Garda Richie Nolan famously summonsed one three-card-trick merchant under the Gaming and Lotteries Act. The defendant subsequently appeared at Enniscorthy District Court, where Richie, in cross-examination of the accused, required him to demonstrate the trick for Judge Donnchadh Ó Buachalla, to see if he could spot the lady. Otherwise, arrests were few.

Along the way, Peter Finn – and his assistant Pat McDonald – had to work closely with Wexford county council and the ploughing association, whose long serving chief, Anna May McHugh, made a strong impression. He described her as 'a very good operator and a forceful woman. There was no messing with Anna May and, at the same time, she was lovely to work with.'

Thanks to Garda Superintendent Peter Finn

45

WIDE OPEN SPACES

> 'Give me land, lots of land'
> COLE PORTER ('DON'T FENCE ME IN')

He dismisses standard Irish corn growing as bad economics and a threat to the environment.

In the past, Jim McCarthy (born 1957) has sown apoplexy through the ranks of the Irish Farmers' Association. He has addressed the farm management association in the UK – and reckons he was lucky to escape alive after airing his views.

Though he has resided at Castledermot in County Kildare since 1981, the Cork accent is as strong as it was when he left Cloyne. This is the man who has controlled up to 2,700 acres of tillage (mainly in Counties Carlow and

Kildare). Yet he does not own so much as one of those acres, and he sold his plough in 1998.

The scale of his activities on rented land was at one stage so large that he was considered by some of his neighbours a threat to all that they held dear. Now his Irish undertaking appears to be of tiny, window-box proportions compared with his latest venture, which has attracted Irish and British investment into three farms in Argentina.

The three farms embrace a total of 31,000 acres — all under tillage, and all looked after by fewer than a dozen workers. The land is flat and featureless, with fences which run for six and a half miles, and 'fields' that are maybe a mile and a half wide.

So, light the blue touchpaper, retire and let Jim McCarthy speak for himself.

On why ploughing is quaint but unnecessary 'Most grains are planted one and a half to two inches deep, but we insist on ploughing ten inches deep. When you plough ten inches deep, you move 2,800 tonnes of soil per acre. The diesel and time consumed are huge. I was looking at this and saying this doesn't make sense, because nature does not move the soil itself. Nature will always keep land covered. If you do anything with the soil, nature's reaction is to cover it up again with green material. It immediately wants to grow something. It will grow lots of things without ploughing it. In Australia, I met a man who started life as a sheep shearer with nothing, and the year I was with him, he and his family had planted 50,000 acres on their own land. And they had never ploughed anything.'

On the traditional view of land in Ireland 'Family farming has survived because of subsidies and a huge contribution of free family labour. As a business model, it's a failed model. It does not work. It's the contribution of free family labour. We have got to be honest and say that family farming can only survive where it is well supported financially by governments – through inflated prices or direct subvention.'

On agriculture in Argentina 'The Argentinians are much better farmers than we are. They have always had low prices on the world market. They have never been subsidised. The famous cattle herds are being displaced, and Argentina will become a beef importer. The Argentinians will tell you that the worst thing the Europeans ever brought to Argentina was the plough. There's about 80 million acres of tillage farm in Argentina – 80 per cent is no-till, conservation agriculture. It makes huge environmental sense. When you walk in a field that's been ploughed, you know the beautiful smell you get from the soil? From freshly turned soil? That's CO_2, the organic matter in the soil being oxidised up into the air as CO_2.'

On farming as a business 'I visited a farm in Russia that was growing 25,000 acres of sugar beet, and the sugar beet was fantastic. They were growing other things as well, and they had taken deliver of sixty new 24-foot New Holland combines. Sixty! The total Irish market is probably sixty combines.'

On the myth that family farming is the only way There are seven or eight investment funds now that five years ago had nothing to do with agriculture. They are all putting at least a billion euro apiece into farming. Farming is the business of the future, but not farming eighty acres, because it won't be commercial. I feel incredibly sorry for traditional family farms because of the peer pressure to remain farming and to be making no money.[2]

Thanks to Jim McCarthy

46

GIVE UP YER 'OUL PLOUGHS?

In 1977, agricultural science graduate Liam Stafford submitted a thesis for his master's degree. His dissertation examined the 'cumbersome' practice of ploughing, its consequences and the alternatives. He pointed to data assembled from ploughed Illinois maize fields where adjacent watercourses were polluted by muck and chemicals. The average soil loss was put at 35.4 hectare per annum.

Liam Stafford pointed to the potential for fuel savings, for the need for less manpower and fertiliser, for better soil-moisture retention, and for a healthier biodiversity by changing to what he described as a 'no-till' system, backed up by the use of herbicides. He was mirroring a trend at the time – a trend that encouraged some Irish farmers to have a go at 'no-till'. The experiment backfired as a stubborn weed called sterile brume dogged the fields. The new approach was rapidly abandoned.

In 2009, the Oireachtas committee on agriculture, fisheries and food, chaired by Deputy Johnny Brady, welcomed a delegation from an organisation called Conservation Agriculture Ireland (CAI) to Leinster House. The no-till subversives were back, and this time stalking the corridors of power.

The CAI brought with them a lurid picture of the town of Clonmel, semi-submerged under the waters of the Suir, and they suggested one way of lessening the likelihood of future repeat inundations. The long-term weather forecast, as a result of global warming, is for more winter storms – and more floods. The delegation, led by Meath based agronomist Gerry Bird and Tipperary's John Geraghty, called for a new approach to farming.

Give up yer oul' ploughs, they argued, and they backed up their case by outlining an alternative. Direct drilling of seed into the ground is the way to go, they reasoned. Do not plough; do not leave good dark earth routinely exposed to the elements; do not pound the ground more than necessary with hefty machinery. Get radical, was their message.

Ploughing, as John Geraghty explained to the assembled politicians, puts oxygen into the soil. The oxygen then burns off the organic matter in that soil. Left alone, that organic matter is capable of holding up to 90 percent of its own weight in water. Water retention is useful not only in dealing with drought but also as a means of heading off floods.

And they went on. Leaving the organic matter, composed of broken-down old plant material, adds to the natural fertility of the fields. Making up the difference with artificial manures only contributes to greenhouse-gas emissions, and the gradual ruination of the planet. 'Unsustainable!' they chorused.

And they went on. Ploughing not only impoverishes the soil, it also leaves land at the mercy of erosion. Repeated ploughing wrecks the soil structure. Direct drilling, on the other hand, encourages biodiversity: more fungi, more worms. Labour costs are reduced. The amount of ozone-destroying diesel required to produce a crop is drastically reduced – from 85 litres per hectare down to 45, according to one trial at Oak Park in Carlow. In short, they echoed all the prescient points made by Liam Stafford thirty-two years earlier.

'Conservation agriculture is now practised as far south as the Falkland Islands, through the Equator, Kenya, Uganda and as far north as Finland,' John Geraghty told the committee. 'It is worked in semi-arid areas, for example in Namibia, to areas of the most intensive and heavy rainfall – instances such as Chile and Brazil.'

He noted that half the arable area in the United Kingdom is now under a non-plough regime. The perplexed speaker compared the UK's total of 50 percent with a figure of less than 3 percent in Ireland. The country has been left behind in the global shift, as non-plough has received little or no official encouragement.

Give up yer oul' ploughs, anyone?

Minutes of Comhchoiste um Thalmhaúocht, Iasciagh and Bia, 29 April 2009

'A Study of the Direct Drilling of Grass', dissertation submitted to the National University of Ireland by Liam Tomás Stafford, B. Agr. Sc., 1977